# Legal Notice

This book is copyright 2014 with all rights reserved. It is illegal to copy, distribute, or create derivative works from this book in whole or in part or to contribute to the copying, distribution, or creating of derivative works of this book.

ISBN-13: 978-1500433642
ISBN-10: 1500433640

**This book is dedicated to all my students over the past 15 years, I have learned just as much from all of you as you have learned from me.**

## BOOKS BY DR. STEVE WARNER FOR COLLEGE BOUND STUDENTS

28 SAT Math Lessons to Improve Your Score in One Month
- Beginner Course
- Intermediate Course
- Advanced Course

320 SAT Math Problems Arranged by Topic and Difficulty Level

320 SAT Math Subject Test Problems Arranged by Topic and Difficulty Level
- Level 1 Test
- Level 2 Test

SAT Prep Book of Advanced Math Problems

The 32 Most Effective SAT Math Strategies

SAT Prep Official Study Guide Math Companion

ACT Prep Red Book – 320 ACT Math Problems with Solutions

## CONNECT WITH DR. STEVE WARNER

www.facebook.com/SATPrepGet800

www.youtube.com/TheSATMathPrep

www.twitter.com/SATPrepGet800

www.linkedin.com/in/DrSteveWarner

www.pinterest.com/SATPrepGet800

plus.google.com/+SteveWarnerPhD

# 320 SAT Math Subject Test Problems arranged by Topic and Difficulty Level

**Level 1**

A Proven Roadmap to
Your First-Choice College

---

Dr. Steve Warner

© 2014, All Rights Reserved
SATPrepGet800.com © 2013

## Table of Contents

# Actions to Complete Before You Read This Book

## 1. Purchase a TI-84 or equivalent calculator

It is recommended that you use a TI-84 or comparable calculator for the SAT Math Subject Test. Answer explanations in this book will always assume you are using such a calculator.

## 2. Take a practice SAT from the Official Guide to get your preliminary SAT math score

Use this score to help you determine the problems you should be focusing on (see page 9 for details).

## 3. Claim your FREE book

Simply visit the following webpage and enter your email address to receive an electronic copy of *The 32 Most Effective SAT Math Strategies* for FREE:

www.thesatmathprep.com/320SATLvL1.html

## 4. 'Like' my Facebook page

This page is updated regularly with SAT prep advice, tips, tricks, strategies, and practice problems. Visit the following webpage and click the 'like' button.

www.facebook.com/SATPrepGet800

# INTRODUCTION
# THE PROPER WAY TO PREPARE

There are many ways that a student can prepare for the SAT Math Level 1 Subject Test. But not all preparation is created equal. I always teach my students the methods that will give them the maximum result with the minimum amount of effort.

The book you are now reading is self-contained. Each problem was carefully created to ensure that you are making the most effective use of your time while preparing for the test. By grouping the problems given here by level and topic I have ensured that you can focus on the types of problems that will be most effective to improving your score.

I recommend beginning SAT Math Subject Test preparation after you have prepared for the SAT Reasoning Test. All of the math preparation you have already completed will be useful when taking the Subject Test. In particular all the strategies that I teach for the SAT Reasoning Test can be used for the Subject Test as well.

You may want to take the SAT Math Subject Test about one or two months after you have taken the SAT for the first time (assuming you have prepared effectively for it). Note that I recommend three to four months of preparation for the SAT. Only one or two additional months is required for the Subject Test because all of the preparation you have already done will still be useful.

If you have not yet decided if you will take the Level 1 or Level 2 test, I strongly recommend you take a College Board practice test of each one to determine which is best for you. Note that the curve on the Level 2 test is much stronger, so even if you feel like you are doing much worse on the Level 2 test, you may wind up with a higher score. If your Level 2 score is less than 50 points lower than your Level 1 score, you may want to go for the Level 2 test instead of the Level 1 test.

## 1. Using this book effectively

- Begin studying at least one month before test day.
- Practice Math Subject Test problems twenty minutes each day.
- Choose a consistent study time and location.

You will retain much more of what you study if you study in short bursts rather than if you try to tackle everything at once. So try to choose about a twenty minute block of time that you will dedicate to the SAT math subject test each day. Make it a habit. The results are well worth this small time commitment.

- Every time you get a question wrong, **mark it off, no matter what your mistake**.
- Begin each study session by first redoing problems from previous study sessions that you have marked off.
- If you get a problem wrong again, **keep it marked off**.

Note that this book often emphasizes solving each problem in more than one way. Please listen to this advice. The same question is not generally repeated on any SAT Subject Test so the important thing is learning as many techniques as possible.

Being able to solve any specific problem is of minimal importance. The more ways you have to solve a single problem, the more prepared you will be to tackle a problem you have never seen before, and the quicker you will be able to solve that problem. Also, if you have multiple methods for solving a single problem, then on the actual test when you "check over" your work you will be able to redo each problem in a different way. This will eliminate all "careless" errors on the actual exam. Note that in this book the quickest solution to any problem will always be marked with an asterisk (*).

## 2. The magical mixture for success

A combination of three components will maximize your SAT Math Subject Test score with the least amount of effort.

- Learning test taking strategies that work specifically for standardized tests.
- Practicing SAT Math Subject Test problems for a small amount of time each day for about three months before the test.
- Taking about two practice tests before test day to make sure you are applying the strategies effectively under timed conditions.

I will discuss each of these three components in a bit more detail.

**Strategy:** The more SAT specific strategies that you know the better off you will be. Throughout this book you will see many strategies being used. Some examples of basic strategies are "plugging in answer choices," "taking guesses," and "picking numbers." Some more advanced strategies include "identifying arithmetic sequences with linear equations," and "recognizing special triangles inside circles." Pay careful attention to as many strategies as possible and try to internalize them. Even if you do not need to use a strategy for that specific problem, you will certainly find it useful for other problems in the future.

**Practice:** The problems given in this book are more than enough to vastly improve your current SAT Math Subject Test score. All you need to do is work on these problems for about ten to twenty minutes each day over a period of one to two months and the final result will far exceed your expectations.

Let me further break this component into two subcomponents – **topic** and **level**.

**Topic:** You want to practice each of the five general math topics given on the SAT Math Subject Test and improve in each independently. The five topics are **Number Theory**, **Algebra and Functions**, **Geometry**, **Probability and Statistics**, and **Trigonometry**. The problem sets in this book are broken into these five topics.

**Level:** You will make the best use of your time by primarily practicing problems that are at and slightly above your current ability level. For example, if you are struggling with Level 2 Geometry problems, then it makes no sense at all to practice Level 5 Geometry problems. Keep working on Level 2 until you are comfortable, and then slowly move up to Level 3. Maybe you should never attempt those Level 5 problems. You can get a score higher than 700 on the Level 1 Subject Test without answering any of them.

**Tests:** You want to take about two practice tests before test day to make sure that you are implementing strategies correctly and using your time wisely under pressure. For this task you should use actual SAT Math Subject Tests such as those found in "The Official SAT Subject Tests in Mathematics Levels 1 and 2 Study Guide." Take one test every few weeks to make sure that you are implementing all the strategies you have learned correctly under timed conditions.

## 3. Practice problems of the appropriate level

In this book SAT Math Subject Test questions have been split into 5 Levels. Roughly speaking, the questions increase in difficulty as you progress from question 1 to question 50. So you can think of the first 10 problems as Level 1, the next 10 as Level 2 and so on.

Keep track of your current ability level so that you know the types of problems you should focus on. If you are still getting most Level 2 Geometry questions wrong, then do not move on to Level 3 Geometry until you start getting more Level 2 Geometry questions right on your own.

If you really want to refine your studying, then you should keep track of your ability level in each of the five major categories of problems:

- **Number Theory**
- **Algebra and Functions**
- **Probability and Statistics**
- **Geometry**
- **Trigonometry**

If you are stronger in Number Theory than Geometry, then it is okay to practice Level 4 Number Theory problems while you continue to practice Level 2 Geometry problems.

## 4. Practice a small amount every day

Ideally you want to practice doing SAT Math Subject Test problems about twenty minutes each day beginning one to two months before the exam. You will retain much more of what you study if you study in short bursts than if you try to tackle everything at once.

The only exception is on a day you do a practice test. You should do at least two practice tests before you take the test. Ideally you should do your practice tests on a Saturday or Sunday morning.

So try to choose about a twenty minute block of time that you will dedicate to practice every night. Make it a habit. The results are well worth this small time commitment.

## 5. Redo the problems you get wrong over and over and over until you get them right

If you get a problem wrong, and never attempt the problem again, then it is extremely unlikely that you will get a similar problem correct if it appears on the SAT Math Subject Test.

Most students will read an explanation of the solution, or have someone explain it to them, and then never look at the problem again. This is *not* how you optimize your score. To be sure that you will get a similar problem correct on the actual test, you must get the problem correct before taking the real test—and without actually remembering the problem.

This means that after getting a problem incorrect, you should go over and understand why you got it wrong, wait at least a few days, then attempt the same problem again. If you get it right you can cross it off your list of problems to review. If you get it wrong, keep revisiting it every few days until you get it right. Your score *does not* improve by getting problems correct. **Your score improves when you learn from your mistakes.**

## 6. Check your answers properly

When you go back to check your earlier answers for careless errors *do not* simply look over your work to try to catch a mistake. This is usually a waste of time. Always redo the problem without looking at any of your previous work. Ideally, you want to use a different method than you used the first time.

For example, if you solved the problem by picking numbers the first time, try to solve it algebraically the second time, or at the very least pick different numbers. If you do not know, or are not comfortable with a different method, then use the same method, but do the problem from the beginning and do not look at your original solution. If your two answers do not match up, then you know that this a problem you need to spend a little more time on to figure out where your error is.

This may seem time consuming, but that's okay. It is better to spend more time checking over a few problems than to rush through a lot of problems and repeat the same mistakes.

## 7. Guess when appropriate

Answering a multiple choice question wrong will result in a 1/4 point penalty. This is to discourage random guessing. If you have no idea how to do a problem, no intuition as to what the correct answer might be, and you cannot even eliminate a single answer choice, then *DO NOT* just take a guess. Omit the question and move on.

If, however, you can eliminate even one answer choice, you should take a guess from the remaining four. You should of course eliminate as many choices as you can before you take your guess.

## 8. Pace yourself

Do not waste your time on a question that is too hard or will take too long. After you have been working on a question for about 30 to 45 seconds you need to make a decision. If you understand the question and think that you can get the answer in another 30 seconds or so, continue to work on the problem. If you still do not know how to do the problem or you are using a technique that is going to take a long time, mark it off and come back to it later if you have time.

If you do not know the correct answer, but you can eliminate at least one answer choice, then take a guess. But you still want to leave open the possibility of coming back to it later. Remember that every problem is worth the same amount. Do not sacrifice problems that you may be able to do by getting hung up on a problem that is too hard for you.

## 9. Attempt the right number of questions

Many students make the mistake of thinking that they have to attempt every single question on the SAT Subject Tests. There is no such rule. In fact, most students will increase their score by *reducing* the number of questions they attempt.

Keep in mind that the questions on the test tend to start out easier in the beginning of the section and get harder as you go. Therefore you should not rush through earlier questions in an attempt to get through the whole test. That said, it is okay to skip several questions that you are stuck on and try a few that appear later on.

Note that although the questions tend to get harder as you go, it is not true that each question is harder than the previous question. For example, it is possible for question 25 to be easier than question 24, and in fact, question 25 can even be easier than question 20. But it is unlikely that question 50 would be easier than question 20.

If you are particularly strong in a certain subject area, then you may want to "seek out" questions from that topic even though they may be more difficult. For example, if you are very strong at number theory problems, but very weak at probability problems, then you may want to try every number theory problem no matter where it appears, and you may want to reduce the number of probability problems you attempt.

## 10. Use your calculator wisely.

- Use a TI-84 or comparable calculator if possible when practicing and during the SAT Subject Test.
- Make sure that your calculator has fresh batteries on test day.
- Make sure your calculator is in degree mode. If you are using a TI-84 (or equivalent) calculator press MODE and on the third line make sure that DEGREE is highlighted. If it is not, scroll down and select it. If possible do not alter this setting until you are finished taking your SAT subject test.

Below are the most important things you should practice on your graphing calculator.

- Practice entering complicated computations in a single step.
- Know when to insert parentheses:
    - Around numerators of fractions
    - Around denominators of fractions
    - Around exponents
    - Whenever you actually see parentheses in the expression

## Examples:

We will substitute a 5 in for $x$ in each of the following examples.

| Expression | Calculator computation |
|---|---|
| $\frac{7x+3}{2x-11}$ | (7*5 + 3)/(2*5 – 11) |
| $(3x-8)^{2x-9}$ | (3*5 – 8)^(2*5 – 9) |

- Clear the screen before using it in a new problem. The big screen allows you to check over your computations easily.
- Press the **ANS** button (**2ND (-)** ) to use your last answer in the next computation.
- Press **2ND ENTER** to bring up your last computation for editing. This is especially useful when you are plugging in answer choices, or guessing and checking.

- You can press **2ND ENTER** over and over again to cycle backwards through all the computations you have ever done.
- Know where the $\sqrt{}$, $\pi$, ^, $e^x$, **LOG** and **LN** buttons are so you can reach them quickly.
- Change a decimal to a fraction by pressing **MATH ENTER ENTER**.
- Press the **MATH** button - in the first menu that appears you can take cube roots and *n*th roots for any *n*. Scroll right to **NUM** and you have **lcm(** and **gcd(**. Scroll right to **PRB** and you have **nPr**, **nCr**, and **!** to compute permutations, combinations and factorials very quickly.
- Know how to use the **SIN**, **COS** and **TAN** buttons as well as **SIN$^{-1}$**, **COS$^{-1}$** and **TAN$^{-1}$**.

You may find the following graphing tools useful.

- Press the **Y=** button to enter a function, and then hit **ZOOM 6** to graph it in a standard window.
- Practice using the **WINDOW** button to adjust the viewing window of your graph.
- Practice using the **TRACE** button to move along the graph and look at some of the points plotted.
- Pressing **2ND TRACE** (which is really **CALC**) will bring up a menu of useful items. For example selecting **ZERO** will tell you where the graph hits the *x*-axis, or equivalently where the function is zero. Selecting **MINIMUM** or **MAXIMUM** can find the vertex of a parabola. Selecting **INTERSECT** will find the point of intersection of 2 graphs.

# PROBLEMS BY LEVEL AND TOPIC WITH FULLY EXPLAINED SOLUTIONS

**Note:** The quickest solution will always be marked with an asterisk (*).

## LEVEL 1: NUMBER THEORY

1. Which of the following numbers is a COUNTEREXAMPLE to the statement "Every positive integer greater than 23 is either prime or divisible by 3, 5, or 7" ?

   (A) 17
   (B) 21
   (C) 25
   (D) 91
   (E) 121

* We are looking for a positive integer greater than 23 that is not prime AND not divisible by 3, 5, or 7.

25 is divisible by 5

91 is divisible by 7

$121 = 11^2$. So 121 is not prime and not divisible by 3, 5, or 7. So the answer is choice (E).

**Note:** We can eliminate 17 and 21 since they are *not* greater than 23.

**Definitions:** The **integers** are the counting numbers together with their negatives.

$$\{\ldots,-4,-3,-2,-1,0,1,2,3,4,\ldots\}$$

The **positive integers** consist of the positive numbers from that set.

$$\{1,2,3,4,\ldots\}$$

A **prime number** is a positive integer that has **exactly** two factors (1 and itself). Here is a list of the first few primes:

$$2, 3, 5, 7, 11, 13, 17, 19, 23, \ldots$$

Note that 1 is **not** prime. It only has one factor!

2. Marco is drawing a time line to represent a 500-year period of time. If he makes the time line 80 inches long and draws it to scale, how many inches will represent each year?

   (A) 0.14
   (B) 0.15
   (C) 0.16
   (D) 0.17
   (E) 0.18

* **Solution by setting up a ratio:** This is a simple ratio. We begin by identifying 2 key words that tell us what 2 things are being compared. In this case, such a pair of key words is "years" and "inches."

| | | |
|---|---|---|
| years | 500 | 1 |
| inches | 80 | $x$ |

Now draw in the division symbols and equal sign, cross multiply and divide the corresponding ratio to find the unknown quantity $x$.

$$\frac{500}{80} = \frac{1}{x}$$

$$500x = (1)(80)$$

$$x = \frac{80}{500} = .16$$

This is choice (C).

3. When six given real numbers are multiplied together, the product is positive. Which of the following could be true about the six numbers?

   (A) One is negative, four are positive, and one is zero.
   (B) Two are negative, three are positive, and one is zero.
   (C) One is negative and five are positive
   (D) Three are negative and three are positive.
   (E) Four are negative and two are positive.

* When we multiply several numbers together, the product will be positive if there is an even number of negative factors, and the rest are positive. The answer choice satisfying this condition is choice (E).

4. If the square root of the cube root of a number is 3, what is the number?

    (A) 729
    (B) 243
    (C) 187
    (D) 81
    (E) 27

**Solution by starting with choice C:** Let's start with choice C and guess that the number is 187. The cube root of 187 is approximately 5.7, and the square root of 5.7 is approximately 2.4. Since this is too small we can eliminate choices C, D, and E.

Let's try choice B next. The cube root of 243 is approximately 6.2, and the square root of 6.2 is approximately 2.5. So we can eliminate choice B, and the answer must be choice (A).

Let's just check that choice A is actually the answer. The cube root of 729 is 9, and the square root of 9 is 3. So the answer is (A).

**Note:** When plugging in or checking answer choices it is a good idea to start with choice C unless there is a specific reason not to. In this problem eliminating choice C allows us to eliminate two more answer choices.

**Algebraic solution using radicals:** If we let $x$ represent the number, then we are given that $\sqrt{\sqrt[3]{x}} = 3$. Squaring each side of this equation gives $\sqrt[3]{x} = 3^2 = 9$. Cubing each side of this last equation gives

$$x = 9^3 = 729.$$

This is choice (A).

***Algebraic solution using fractional exponents:** If we let $x$ represent the number, then we are given that $\left(x^{\frac{1}{3}}\right)^{\frac{1}{2}} = 3$ (See the tenth law in the table below). So we have $x^{\frac{1}{6}} = 3$ (See the fifth law in the table below). We now raise each side of this last equation to the sixth power to get

$$x = 3^6 = 729.$$

This is choice (A).

## Laws of Exponents

| Law | Example |
|---|---|
| $x^0 = 1$ | $3^0 = 1$ |
| $x^1 = x$ | $9^1 = 9$ |
| $x^a x^b = x^{a+b}$ | $x^3 x^5 = x^8$ |
| $x^a/x^b = x^{a-b}$ | $x^{11}/x^4 = x^7$ |
| $(x^a)^b = x^{ab}$ | $(x^5)^3 = x^{15}$ |
| $(xy)^a = x^a y^a$ | $(xy)^4 = x^4 y^4$ |
| $(x/y)^a = x^a/y^a$ | $(x/y)^6 = x^6/y^6$ |
| $x^{-1} = 1/x$ | $3^{-1} = 1/3$ |
| $x^{-a} = 1/x^a$ | $9^{-2} = 1/81$ |
| $x^{1/n} = \sqrt[n]{x}$ | $x^{1/3} = \sqrt[3]{x}$ |
| $x^{m/n} = \sqrt[n]{x^m} = (\sqrt[n]{x})^m$ | $x^{9/2} = \sqrt{x^9} = (\sqrt{x})^9$ |

# LEVEL 1: ALGEBRA AND FUNCTIONS

5. If $y = 2x^3 - 3.2$, for what value of $x$ is $y = 5$ ?

   (A) 1.6
   (B) 1.9
   (C) 2.4
   (D) 3.5
   (E) 4.1

**Solution by starting with choice C:** We start with choice C and take a guess that $x = 2.4$. We then have $y = 2(2.4) \wedge 3 - 3.2 = 24.448$ (just use your calculator). This is too big. So we can eliminate choices C, D, and E.

Let's try choice B next and guess that $x = 1.9$. Then

$$y = 2(1.9) \wedge 3 - 3.2 = 10.518.$$

This is still too big. So we can eliminate choice B and the answer must be choice (A).

Let's just verify that choice A does actually give the correct answer. If $x = 1.6$, then $y = 2(1.6) \wedge 3 - 3.2 = 4.992$. This is close enough. So the answer is in fact choice (A).

* **Algebraic solution:** Let's substitute 5 in for $y$ to get $5 = 2x^3 - 3.2$.

We now add 3.2 to each side of this equation to get $8.2 = 2x^3$.

Now divide each side of this equation by 2 to get $4.1 = x^3$

Finally, take the cube root of each side of this equation (use your calculator) to get $x \approx 1.6$, choice (A).

6. If $3a + ab = 56$ and $b - 8 = -4$, what is the value of $a$ ?

(A) 2
(B) 4
(C) 6
(D) 8
(E) 10

* **Algebraic solution:** We start by solving the second equation for $b$ to get $b = 4$. We then substitute $b = 4$ into the first equation to get

$$3a + 4a = 56.$$

So we have $7a = 56$, and therefore $a = 8$, choice (D).

**Notes:** (1) We can solve the equation $b - 8 = -4$ informally by asking "what number minus 8 gives us $-4$?" Well, this number is 4.

(2) If you prefer to solve the equation $b - 8 = -4$ formally, simply add 8 to each side of the equation.

(3) Similarly, we can solve the equation $7a = 56$ informally by asking "7 times what number is 56?" Well, this number is 8.

(4) If you prefer to solve the equation $7a = 56$ formally, simply divide each side of the equation by 7.

7. If $f(x) = \frac{3}{x^2}$ for $x > 0$, then $f(2.5) =$

(A) $\frac{8}{25}$

(B) $\frac{2}{5}$

(C) $\frac{12}{25}$

(D) $\frac{3}{5}$

(E) $\frac{18}{25}$

* $f(2.5) = \frac{3}{2.5^2} = \frac{12}{25}$, choice (C).

**Notes:** (1) We can do this computation right in our calculator:

3 / 2.5 ^ 2 ENTER

The output will be .48

(2) We can change .48 to a fraction in our TI-84 calculator by pressing MATH ENTER ENTER.

Alternatively, we can change each of the fractions in the answer choices to decimals by performing the appropriate division. For example, to change $\frac{12}{25}$ to a decimal we simply divide 12 by 25 in our calculator.

8. If $3x - 7x = 2x - 8x + 6$, then $x =$

(A) 0
(B) 1
(C) 2
(D) 3
(E) 4

* **Algebraic solution:** We start by combing like terms on each side of the equation to get $-4x = -6x + 6$. We now add $6x$ to each side of this last equation to get $2x = 6$. Finally, we divide by 2 to get $x = 3$, choice (D).

**Remark:** We can also solve this problem by starting with choice C. I leave the details of this solution to the reader.

9. If $2x^2 - 11 = 5 - 2x^2$, what are all possible values of $x$ ?

(A) 2 only
(B) $-2$ only
(C) 0 only
(D) 2 and $-2$ only
(E) 0, 2, and $-2$

**Solution by plugging in the answer choices:** According to the answer choices we need only check 0, 2, and $-2$.

$x = 0$: $2(0)^2 - 11 = 5 - 2(0)^2$ $-11 = 5$ False

$x = 2$: $2(2)^2 - 11 = 5 - 2(2)^2$ $-3 = -3$ True

$x = -2$: $2(-2)^2 - 11 = 5 - 2(-2)^2$ $-3 = -3$ True

So the answer is choice (D).

**Note:** Since all powers of $x$ in the given equation are even, 2 and $-2$ must give the same answer. So we didn't really need to check $-2$.

* **Algebraic solution:** We add $2x^2$ to each side of the given equation to get $4x^2 - 11 = 5$. We then add 11 to get $4x^2 = 16$. Dividing each side of this last equation by 4 gives $x^2 = 4$. We now use the **square root property** to get $x = \pm 2$. So the answer is choice (D).

**Notes:** (1) The equation $x^2 = 4$ has two solutions: $x = 2$ and $x = -2$. A common mistake is to forget about the negative solution.

(2) The **square root property** says that if $x^2 = c$, then $x = \pm\sqrt{c}$.

This is different from taking the positive square root of a number. For example, $\sqrt{4} = 2$, whereas the equation $x^2 = 4$ has two solutions $x = \pm 2$.

(3) Another way to solve the equation $x^2 = 4$ is to subtract 4 from each side of the equation, and then factor the difference of two squares as follows:

$$x^2 - 4 = 0$$
$$(x-2)(x+2) = 0$$

We now set each factor equal to 0 to get $x - 2 = 0$ or $x + 2 = 0$.

So $x = 2$ or $x = -2$.

10. If $x^2 - y^2 = 36$ and $x - y = -9$, then $x + y =$

    (A) −9
    (B) −4
    (C) 0
    (D) 4
    (E) 9

* Substituting into the formula $(x + y)(x - y) = x^2 - y^2$ we have

$$(x + y)(-9) = 36.$$

It follows that $x + y = \frac{36}{-9} = -4$, choice (B).

11. If $a^{5c-2} = a^{3c+7}$ for all real values of $a$, what is the value of $c$ ?

    (A) 2.5
    (B) 3
    (C) 3.5
    (D) 4
    (E) 4.5

* **Algebraic solution:** Since the bases are the same we set the exponents equal to each other. So we have $5c - 2 = 3c + 7$. We subtract $3c$ from each side of this equation to get $2c - 2 = 7$. We now add 2 to each side of this last equation to get $2c = 9$. Finally, $c = \frac{9}{2} = 4.5$, choice (E).

**Notes:** (1) Here we used the fact that if $a^x = a^y$, then $x = y$.

(2) This problem can also be solved by plugging in the answer choices. I leave the details to the reader.

12. Which of the following graphs could not be the graph of a function?

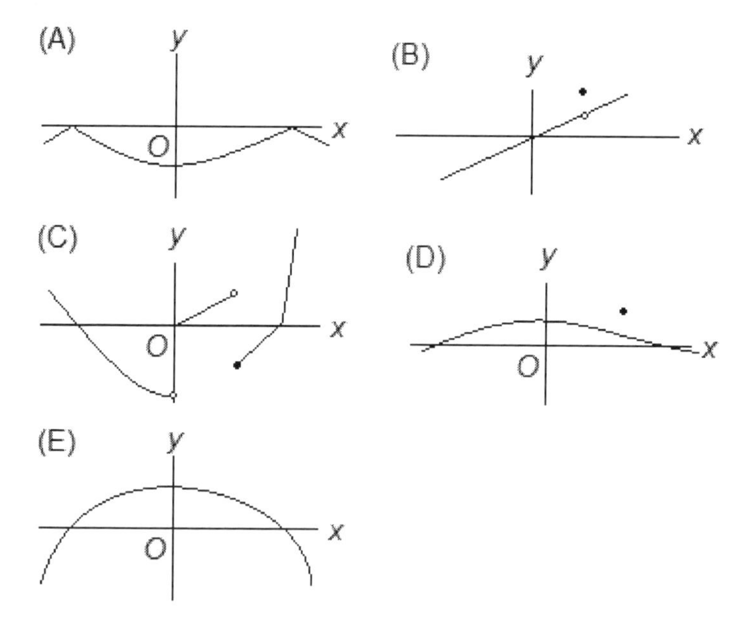


* Only choice (D) fails the **vertical line test**. In other words, we can draw a vertical line that hits the graph more than once:

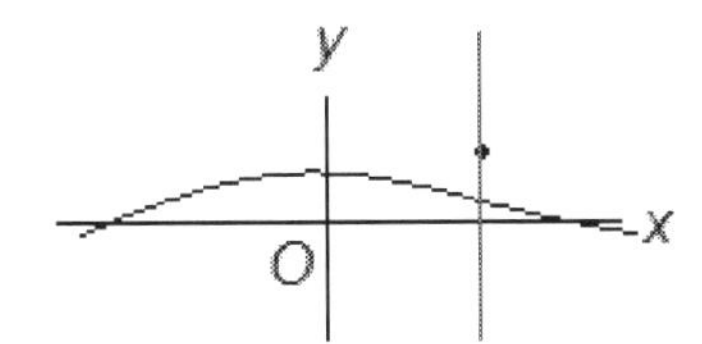


So the answer is choice (D).

13. If $7x + y = 6$ and $5x + y = 2$, what is the value of $6x + y$ ?

(A) $-8$
(B) 4
(C) 6
(D) 12
(E) 18

*** Solution by adding the two equations:** We add the two equations:

$$\begin{aligned} 7x + \ y &= 6 \\ \underline{5x + \ y} &\underline{= 2} \\ 12x + 2y &= 8 \end{aligned}$$

Now observe that $12x + 2y = 2(6x + y)$. So $6x + y = \frac{8}{2} = 4$, choice (B).

**Solution by subtracting the equations:** We subtract the two equations to isolate $x$.

$$\begin{aligned} 7x + \ y &= 6 \\ \underline{5x + \ y} &\underline{= 2} \\ 2x \qquad &= 4 \end{aligned}$$

So $x = 2$. Substituting $x = 2$ back into the first equation we get

$$\begin{aligned} 7(2) + y &= 6 \\ 14 \ + y &= 6 \\ y &= -8 \end{aligned}$$

So, $6x + y = 6(2) - 8 = 12 - 8 = 4$, choice (B).

14. If $|4 - 7x| > 30$, which of the following is a possible value of $x$?

(A) $-4$
(B) $-2$
(C) 2
(D) 3
(E) 4

**Solution by starting with choice C:** Let's start with choice (C) and guess that $x = 2$. Then $|4 - 7x| = |4 - 7 \cdot 2| = |4 - 14| = |-10| = 10$. Since 10 is not greater than 30 we can eliminate choice (C). A moment's thought may lead you to suspect choice (A) (if you do not see this it is okay – just keep trying the answer choices until you get to it). Now, setting $x = -4$ gives us $|4 - 7x| = |4 - 7(-4)| = |4 + 28| = |32| = 32$. Since 32 is greater than 30, the answer is choice (A).

* **Partial algebraic solution:** We can try to simply eliminate the absolute values and solve the resulting inequality.

$$4 - 7x > 30$$
$$-7x > 26$$
$$x < \frac{26}{-7} \sim -3.714$$

Since $-4 < -3.714$, the answer is choice (A).

**Note:** The inequality changed direction in the last step because we divided each side of the inequality by a negative number.

**Complete algebraic solution:** The given absolute value inequality is equivalent to $4 - 7x < -30$ or $4 - 7x > 30$. Let's solve these two inequalities.

$$4 - 7x < -30 \quad \text{or} \quad 4 - 7x > 30$$
$$-7x < -34 \quad \text{or} \quad -7x > 26$$
$$x > \frac{34}{7} \sim 4.857 \quad \text{or} \quad x < \frac{26}{-7} \sim -3.714$$

So $x < -3.714$ or $x > 4.857$. Since $-4 < -3.714$, the answer is choice (A).

# LEVEL 1: GEOMETRY

15. If the point $(3, y)$ is the intersection of the graphs of $x^2 = 9$ and $x = \frac{8y^3}{9}$, then $y =$

(A) $-3$

(B) $-\frac{3}{2}$

(C) 1

(D) $\frac{3}{2}$

(E) 2

* A point of intersection of two graphs lies on both graphs. In particular, the point $(3, y)$ lies on the graph of $x = \frac{8y^3}{9}$, and it follows that $3 = \frac{8y^3}{9}$. We multiply this equation by $\frac{9}{8}$ to get $y^3 = \frac{27}{8}$. So $y = \frac{3}{2}$, choice (D).

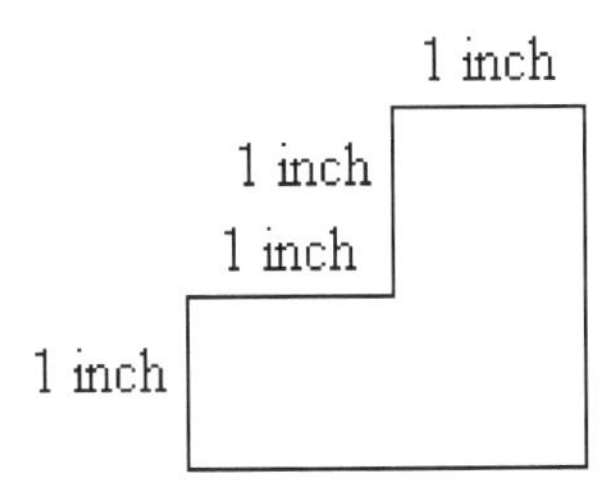


16. How many figures of the size and shape above are needed to completely cover a rectangle measuring 60 inches by 40 inches?

    (A) 200
    (B) 400
    (C) 600
    (D) 800
    (E) 1000

* **Solution by dividing areas:** The area of the given figure is 3 square inches and the area of the rectangle is $60 \cdot 40 = 2400$ square inches. We can see how many of the given figures cover the rectangle by dividing the two areas.

$$\frac{2400}{3} = 800, \text{ choice (D).}$$

**Note:** We can get the area of the given figure by splitting it into 3 squares each with area 1 inch$^2$ as shown below. Then $1 + 1 + 1 = 3$.

1 inch
1 inch
1 inch
1 inch

Another way to get the area of the given figure is to think of it as lying inside a square of side length 2 inches as shown below.

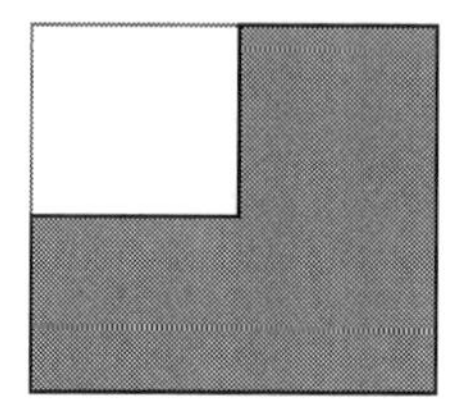

The area of the big square is $2 \cdot 2 = 4$ square inches, and the area of the small square is $1 \cdot 1 = 1$ square inch. Therefore the area of the given figure is $4 - 1 = 3$ square inches.

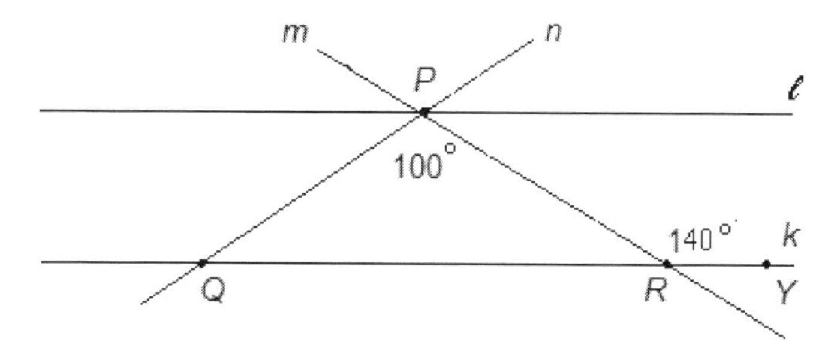


17. In the figure above, line $\ell$ is parallel to line $k$. Transversals $m$ and $n$ intersect at point $P$ on $\ell$ and intersect $k$ at points $R$ and $Q$, respectively. Point $Y$ is on $k$, the measure of $\angle PRY$ is 140°, and the measure of $\angle QPR$ is 100°. How many of the angles formed by rays $\ell$, $k$, $m$, and $n$ have measure 40° ?

    (A) 4
    (B) 6
    (C) 8
    (D) 10
    (E) 12

* $\angle QRP$ is supplementary with $\angle PRY$. So $m\angle QRP$ is $180 - 140 = 40°$. We can then use vertical angles to get the remaining angles in the lower right hand corner.

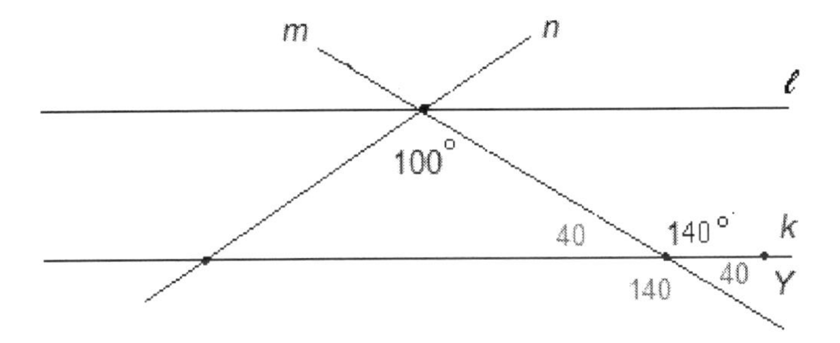


We now use the fact that the sum of the angle measures in a triangle is $180°$ to get that the measure of the third angle of the triangle is $180 - 100 - 40 = 40°$. We then once again use supplementary and vertical angles to get the remaining angles in the lower left hand corner.

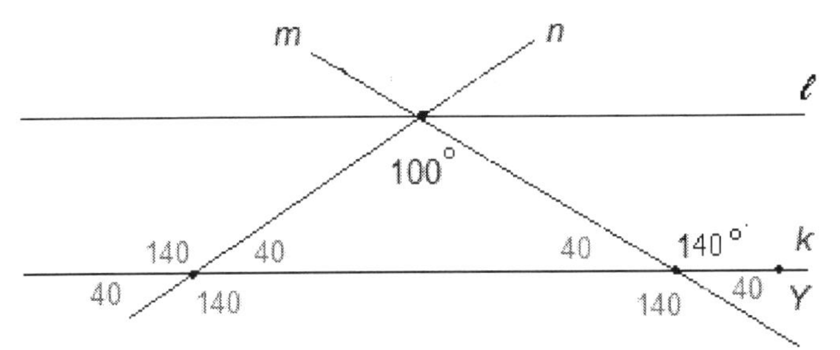


Now notice the following alternate interior angles.

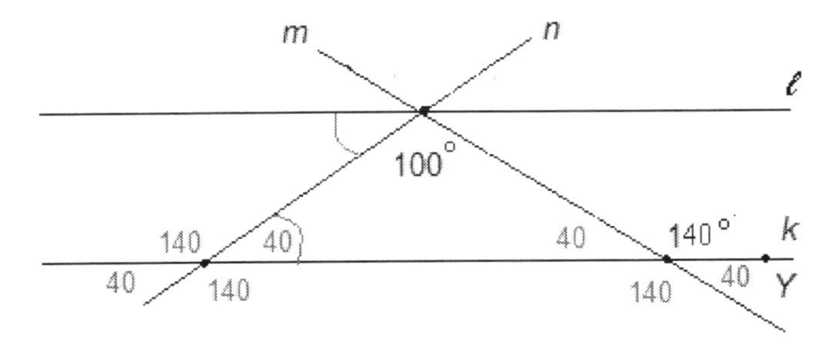


Since alternate interior angles are congruent, we see that the angle marked above has a measure of $40°$. We use supplementary and vertical angles to find the remaining angle measures.

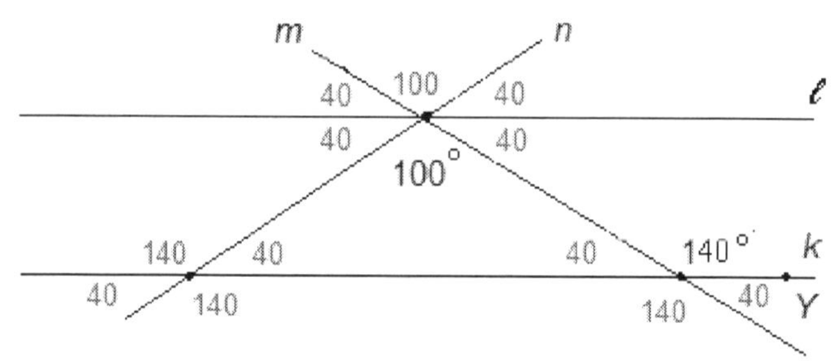


Finally, we see that there are eight angles with measure 40°, choice (C).

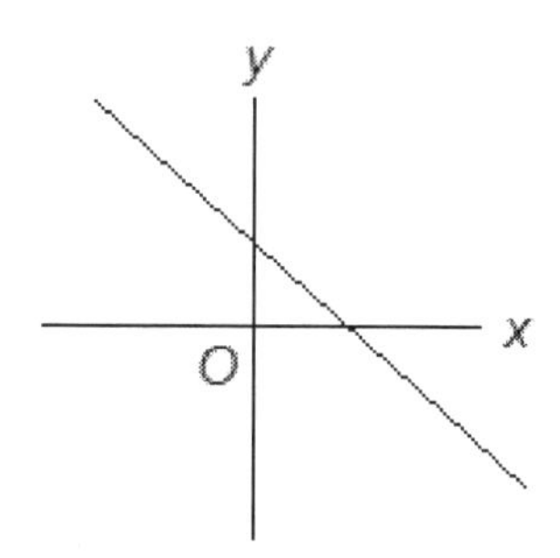


18. The figure above shows the graph of the linear function $f(x) = mx + b$. Which of the following is true about $m$ and $b$ ?

    (A) $m > 0$ and $b > 0$
    (B) $m > 0$ and $b < 0$
    (C) $m < 0$ and $b > 0$
    (D) $m < 0$ and $b < 0$
    (E) $m = 0$ and $b > 0$

* Since the graph is going downwards from left to right, $m < 0$. Since the graph hits the $y$-axis above the $x$-axis, $b > 0$. So the answer is (C).

**Slope formula and linear equations:**

$$\text{Slope} = m = \frac{rise}{run} = \frac{y_2 - y_1}{x_2 - x_1}$$

**Note:** Lines with positive slope have graphs that go upwards from left to right. Lines with negative slope have graphs that go downwards from left to right. If the slope of a line is zero, it is horizontal. Vertical lines have **no** slope, or **undefined** slope (this is different from zero slope).

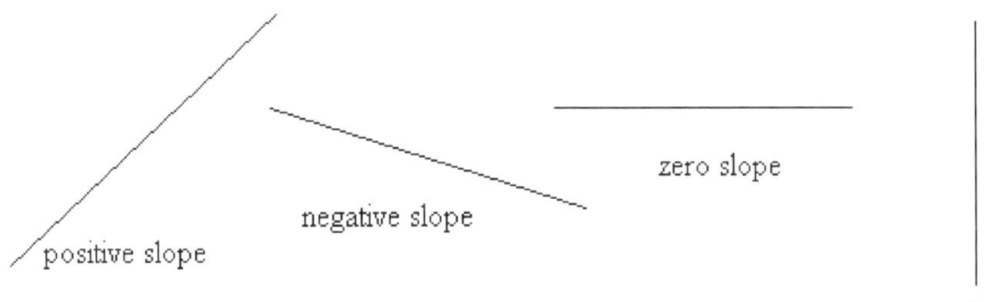


The **slope-intercept form of an equation of a line** is $\boldsymbol{y = mx + b}$ where $m$ is the slope of the line and $b$ is the $y$-coordinate of the $y$-intercept, i.e. the point $(0, b)$ is on the line. Note that this point lies on the $y$-axis.

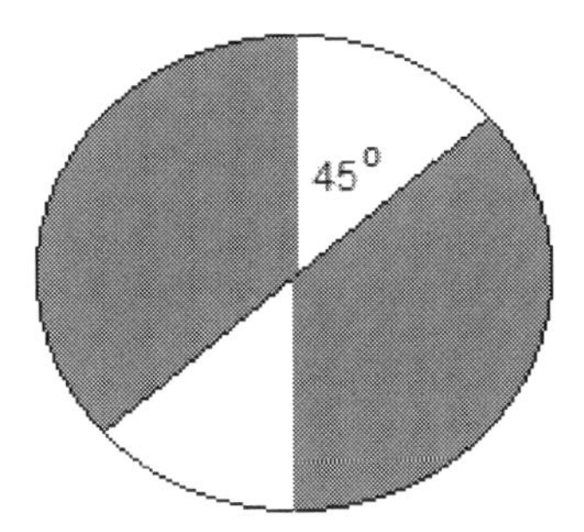


19. In the figure above, what percentage of the circle is shaded?

(A) 65%
(B) 70%
(C) 75%
(D) 80%
(E) 85%

* **Solution by setting up a ratio:** From the figure, the unshaded portion of the circle represents 2(45) = 90° of the circle. Since there are 360° in a circle, the shaded portion represents 360 – 90 = 270° of the circle. So we have $\frac{270}{360} = 0.75 = 75\%$ of the circle is shaded, choice (C).

**Note:** To change a decimal to a percent, multiply by 100, or equivalently move the decimal point two places to the right.

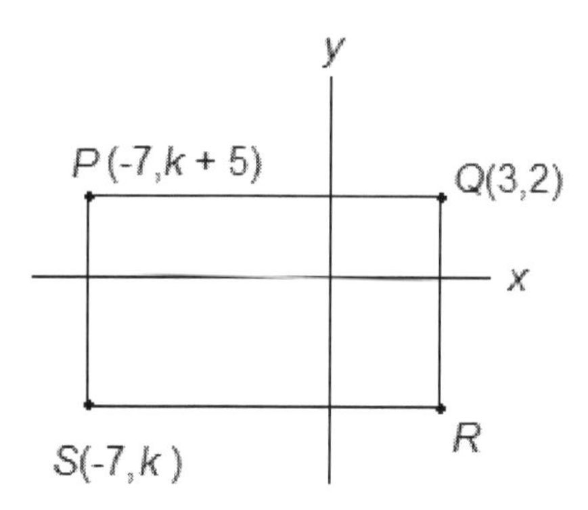


20. In rectangle $PQRS$ above, what are the coordinates of vertex $R$.

(A) (2, –3)
(B) (3, –2)
(C) (3, –2)
(D) (3, –3)
(E) (3, –4)

* First note that $P$ and $Q$ have the same $y$-coordinate so that $k+5=2$. It follows that $k=2-5=-3$.

Now note that $R$ has the same $y$-coordinate as $S$, so that the $y$-coordinate of $R$ is $k=-3$.

Finally note that $R$ has the same $x$-coordinate as $Q$, so the $x$-coordinate of $R$ is 3. Therefore the answer is choice (D).

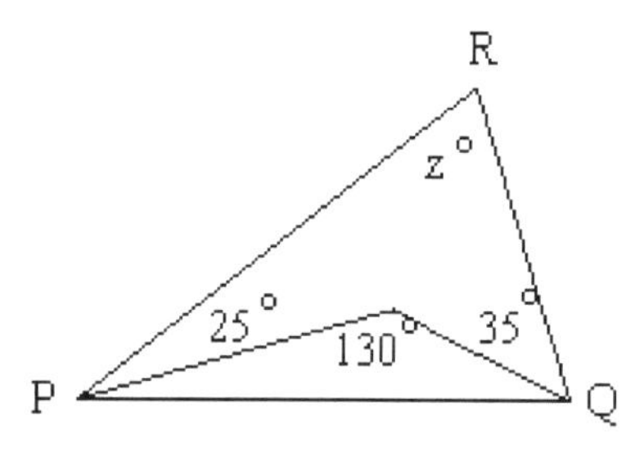


Note: Figure not drawn to scale.

21. In triangle $PQR$ above, what is the value of $z$ ?

(A) 65
(B) 70
(C) 75
(D) 80
(E) 85

**Solution by picking numbers:** Since every triangle has 180 degrees we choose values for the angle measures of the small triangle that add up to $180 - 130 = 50$, say 25 and 25.

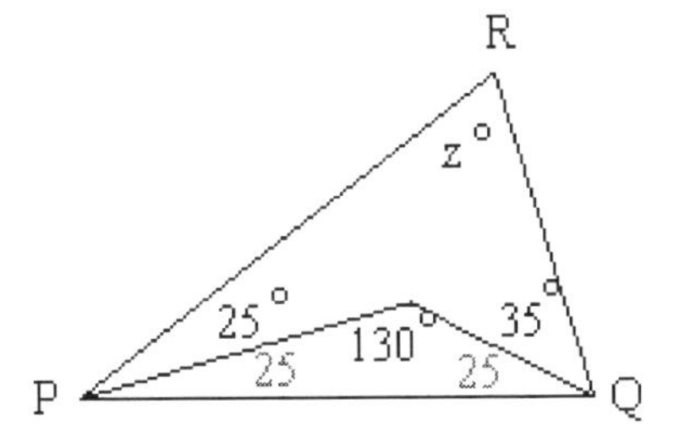


Note: Figure not drawn to scale.

We now see that $z = 180 - 25 - 25 - 35 - 25 = 70$, choice (B).

**Remark:** We could have chosen **any** two numbers that add up to 50 for the angles of the small triangle.

* **Geometric solution:** The two unlabeled angles in the smaller triangle must add up to 50. Therefore

$$z = 180 - 25 - 35 - 50 = 70\text{, choice (B).}$$

22. Right triangle $PQR$ has a hypotenuse of length 11 and one of its legs as length 8. How many possible values are there for the area of triangle $PQR$ ?

    (A) one
    (B) two
    (C) three
    (D) infinitely many
    (E) no such triangle exists

* **Geometric solution:** By the Pythagorean Theorem, the second leg of the right triangle is uniquely determined. The base and height of a right triangle are the two legs. Therefore the area of the triangle is also uniquely determined, and so the answer is choice (A).

**Note:** There is no need to actually compute the area of the triangle to solve this problem.

**Solution by drawing a picture:** Let's draw a picture:

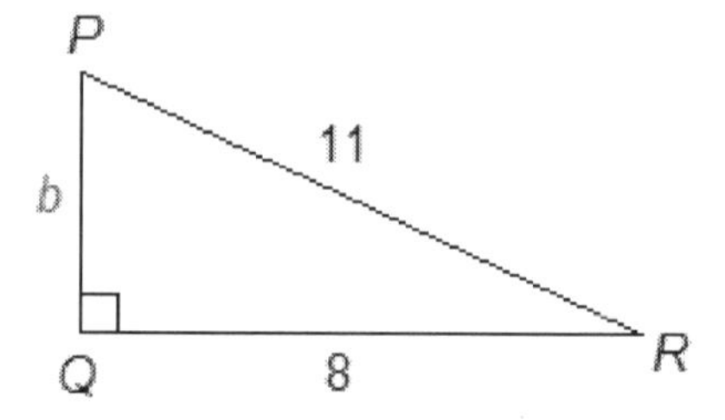


We can now find $b$ by using the Pythagorean Theorem.

$$8^2 + b^2 = 11^2$$
$$64 + b^2 = 121$$
$$b^2 = 57$$
$$b = \sqrt{57}$$

So the area of the triangle is $A = \frac{1}{2} \cdot 8\sqrt{57} = 4\sqrt{57}$.

In particular, there is exactly one possible value for the area of the triangle, choice (A).

**Note:** The equation $b^2 = 57$ would normally have two solutions: $b = \sqrt{57}$ and $b = -\sqrt{57}$. But the length of a side of a triangle cannot be negative, so we reject $-\sqrt{57}$.

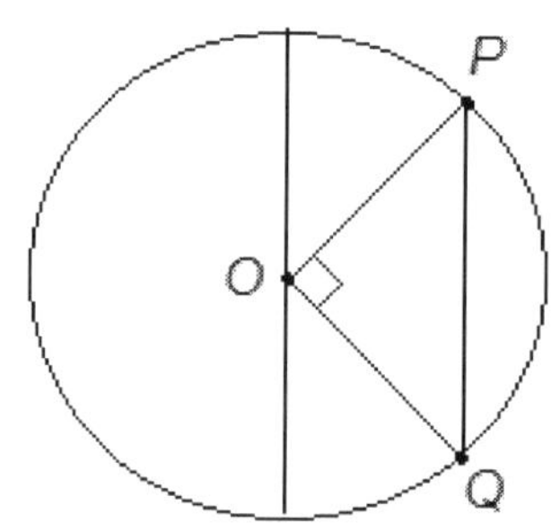


23. The circle shown above has center $O$ and radius 7. What is the length of chord $PQ$?

(A) $\frac{7}{2}$

(B) $\frac{7\sqrt{2}}{2}$

(C) 7

(D) $7\sqrt{2}$

(E) $7\sqrt{3}$

* **Solution using a 45, 45, 90 triangle:** Since the two legs of the right triangle are both radii of the circle, the triangle is an isosceles right triangle, or equivalently a 45, 45, 90 right triangle. It follows that $PQ$ has length $7\sqrt{2}$, choice (D).

**Notes:** (1) A triangle is **isosceles** if it has two sides of equal length. Equivalently, an isosceles triangle has two angles of equal measure.

(2) For the SAT Math Subject Test, it is worth knowing the following two special triangles:

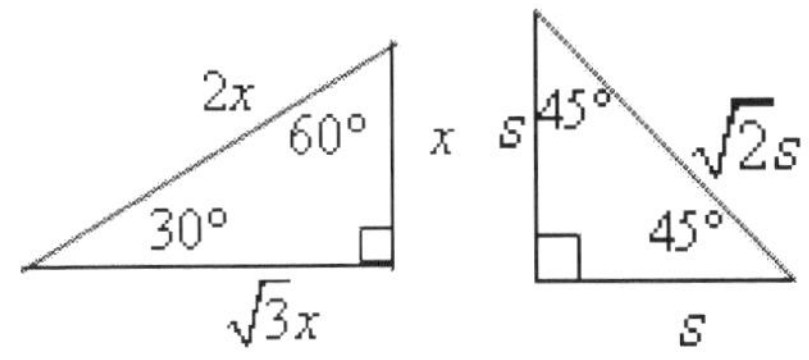


Some students get a bit confused because there are variables in these pictures. But the pictures become simplified if we substitute a 1 in for the variables. Then the sides of the 30, 60, 90 triangle are 1, 2 and $\sqrt{3}$ and the sides of the 45, 45, 90 triangle are 1, 1 and $\sqrt{2}$. The variable just tells us that if we multiply one of these sides by a number, then we have to multiply the other two sides by the same number.

For example, instead of 1, 1 and $\sqrt{2}$, we can have 3, 3 and $3\sqrt{2}$ (here we have $s = 3$), or $\sqrt{2}$, $\sqrt{2}$, and 2 (here $s = \sqrt{2}$). For this problem, we are using the 30, 60, 90 right triangle and $x = 7$.

(3) The hypotenuse of the right triangle can also be found using the Pythagorean Theorem. I leave the details of this solution to the reader.

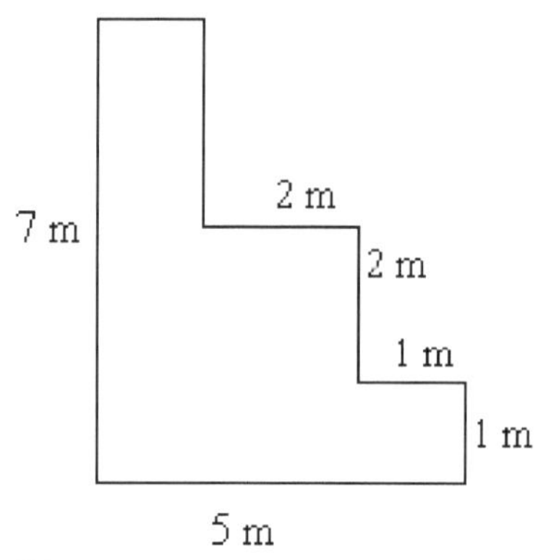

Note: Figure not drawn to scale.

24. What is the area of the figure above?

(A) 15 $m^2$
(B) 17 $m^2$
(C) 19 $m^2$
(D) 21 $m^2$
(E) 23 $m^2$

* We break the figure up into 3 rectangles and compute the length and width of each rectangle.

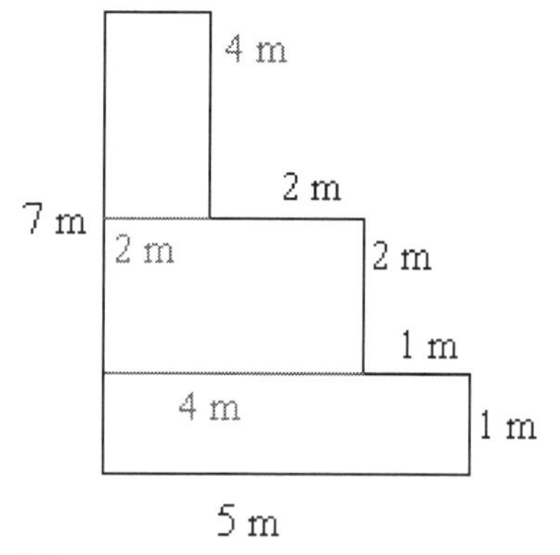

Note: Figure not drawn to scale.

The length and width of the bottom rectangle are 5 and 1 making the area $5 \cdot 1 = 5$ $m^2$.

The length of the middle rectangle is $5 - 1 = 4$, and the width is given as 2. Thus, the area is $4 \cdot 2 = 8$ $m^2$.

The length of the top rectangle is $4 - 2 = 2$, and the width is $7 - 1 - 2 = 4$. Thus, the area is $2 \cdot 4 = 8$ m$^2$.

We then get the total area by adding up the areas of the three rectangles: $5 + 8 + 8 = 21$ m$^2$, choice (D).

**Remark:** Notice that if we have the full length of a line segment, and one partial length of the same line segment, then we get the other length by subtracting the two given lengths.

## LEVEL 1: PROBABILITY AND STATISTICS

25. Of the marbles in a jar, 14 are orange. Sarah randomly takes one marble out of the jar. If the probability is $\frac{2}{7}$ that the marble she chooses is orange, how many marbles are in the jar?

    (A) 4
    (B) 14
    (C) 28
    (D) 49
    (E) 98

**Solution by plugging in answer choices:** Let's start with choice C and guess that there are 28 marbles in the jar. We have $\frac{2}{7} \cdot 28 = 8$. This is too small. So we can eliminate choices A, B, and C.

Let's try choice D next and guess that there are 49 marbles in the jar. We have $\frac{2}{7} \cdot 49 = 14$. This is correct. So the answer is choice (D).

**Remarks:** (1) Saying that the probability is $\frac{2}{7}$ that an orange marble will be chosen is equivalent to saying that $\frac{2}{7}$ of the marbles are orange. So for example, when we guess that there are 49 marbles in the jar, we need to compute $\frac{2}{7}$ of 49. The word "of" always means multiplication.

(2) All of the computations above can be done by hand or in your calculator. For example, to compute $\frac{2}{7} \cdot 28$, simply type 2 / 7 * 28 followed by ENTER. The output will be 8.

* **Algebraic solution:** Let $x$ be the total number of marbles in the jar. We are given that $\frac{2}{7}x = 14$. We multiply each side of this equation by $\frac{7}{2}$ to get that $x = 14 \cdot \frac{7}{2} = 49$, choice (D).

26. Between Town A and Town B there are 5 roads, between Town B and Town C there are 2 roads, and between Town C and Town D there are 4 roads. If a traveler were to travel from Town A to Town D, passing first through B, then through C, how many different routes does he have to choose from?

    (A) 11
    (B) 20
    (C) 40
    (D) 60
    (E) 80

* **Solution using the counting principle:** The counting principle says that when you perform events in succession you multiply the number of possibilities. So we get $5 \cdot 2 \cdot 4 = 40$ different routes, choice (C).

**Remark:** The **counting principle** says that if one event is followed by a second independent event, the number of possibilities is multiplied.

More generally, if $E_1, E_2, \ldots, E_n$ are $n$ independent events with $m_1, m_2, \ldots, m_n$ possibilities, respectively, then event $E_1$ followed by event $E_2$, followed by event $E_3$, ..., followed by event $E_n$ has $m_1 \cdot m_2 \cdots m_n$ possibilities.

In this question there are 3 events: "choosing a road between Town A and Town B," "choosing a road between Town B and Town C," and "choosing a road between Town C and Town D."

27. Kenneth's test average after 7 tests was 81. His score on the 8th test was 92. If all 8 tests were equally weighted, which of the following is closest to his test average after 8 tests?

    (A) 80
    (B) 82
    (C) 84
    (D) 86
    (E) 88

* **Solution by changing the average to a sum:** We change the average to a sum using the formula

$$\text{Sum} = \text{Average} \cdot \text{Number}$$

We are averaging seven numbers. Thus, the **Number** is 7. The **Average** is given to be 81. So the **Sum** of the seven numbers is $81 \cdot 7 = 567$.

When we add 92, the **Sum** becomes $567 + 92 = 659$. The **Average** is then $\frac{659}{8} = 82.375$. This is closest to 82, choice (B).

28. Nine different books are to be stacked in a pile. One book is chosen for the bottom of the pile and another book is chosen for the top of the pile. In how many different orders can the remaining books be placed on the stack?

    (A) 36
    (B) 72
    (C) 5040
    (D) 40,320
    (E) 362,880

* Once we have chosen a book for the bottom and a book for the top of the pile there are seven books left to stack. Thus, it follows that there are $7! = (7)(6)(5)(4)(3)(2)(1) = 5040$ ways to stack these books, choice (C).

**Calculator remark:** You can compute 7! in your calculator as follows. After typing 7, simply press MATH, scroll to PRB and then select ! (or press 4)

# LEVEL 1: TRIGONOMETRY

29. Let $x = \cos\theta$ and $y = \sin\theta$ for any real value $\theta$. Then $x^2 + y^2 =$

    (A) $-1$
    (B) 0
    (C) 1
    (D) 2
    (E) It cannot be determined from the information given

* **Solution using a Pythagorean identity:**

$$x^2 + y^2 = (\cos\theta)^2 + (\sin\theta)^2 = 1$$

This is choice (C).

**Notes:** (1) $(\cos\theta)^2$ is usually abbreviated as $\cos^2\theta$.

Similarly, $(\sin\theta)^2$ is usually abbreviated as $\sin^2\theta$.

In particular, $(\cos\theta)^2 + (\sin\theta)^2$ would be written as $\cos^2\theta + \sin^2\theta$.

(2) One of the most important trigonometric identities is the Pythagorean Identity which says

$$\cos^2 x + \sin^2 x = 1.$$

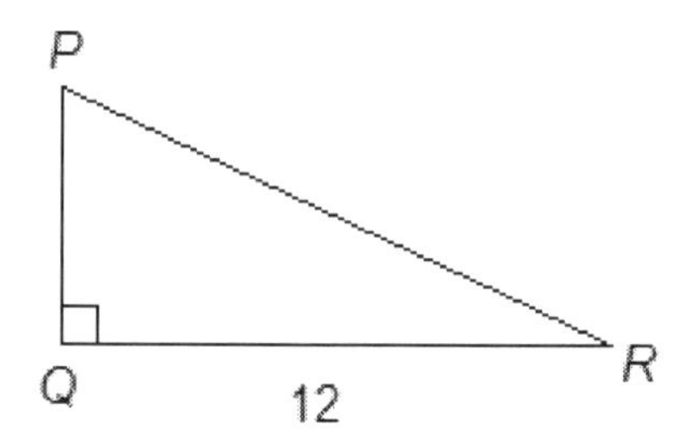


30. In $\Delta PQR$ above, $\tan R = \frac{5}{12}$. What is the length of side $PR$ ?

(A) 11
(B) 13
(C) 15
(D) 16
(E) 17

* Since $\tan R = \frac{\text{OPP}}{\text{ADJ}}$, we have $\frac{5}{12} = \frac{\text{OPP}}{\text{ADJ}}$. Since the adjacent side is 12, the opposite side must be 5. So we have the following picture.

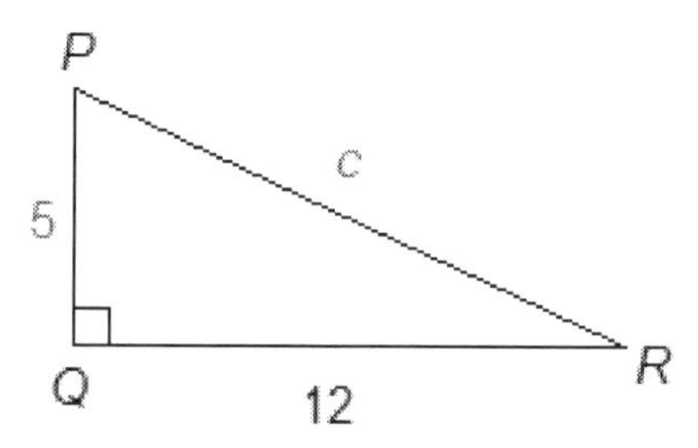


We now find $PR$ by using the Pythagorean Theorem, or better yet, recognizing the Pythagorean triple 5, 12, 13.

So $PR = 13$, choice (B).

**Remarks:** (1) The most common Pythagorean triples are 3,4,5 and 5, 12, 13. Two others that may come up are 8, 15, 17 and 7, 24, 25.

(2) If you don't remember the Pythagorean triple 5, 12, 13, you can use the Pythagorean Theorem which says that if a right triangle has legs of length $a$ and $b$, and a hypotenuse of length $c$, then $c^2 = a^2 + b^2$.

In this problem we have $c^2 = 5^2 + 12^2 = 169$. So $c = 13$.

(3) The equation $c^2 = 169$ would normally have two solutions: $c = 13$ and $c = -13$. But the length of a side of a triangle cannot be negative, so we reject $-13$.

A quick lesson in **right triangle trigonometry** for those of you that have forgotten.

Let's begin by focusing on angle $A$ in the following picture:

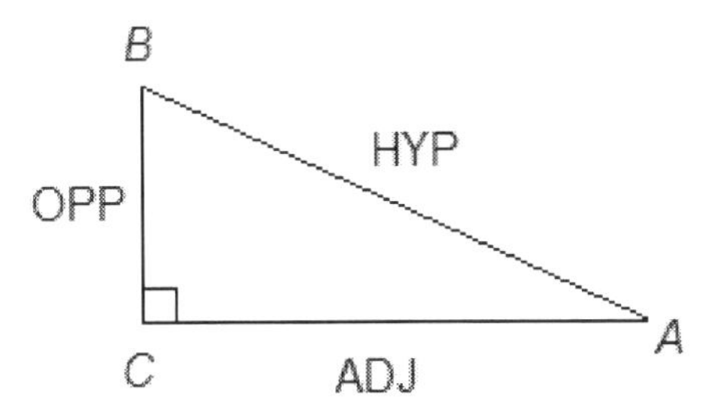


Note that the **hypotenuse** is ALWAYS the side opposite the right angle.

The other two sides of the right triangle, called the **legs**, depend on which angle is chosen. In this picture we chose to focus on angle $A$. Therefore the opposite side is $BC$, and the adjacent side is $AC$.

Now you should simply memorize how to compute the six trig functions:

$$\sin A = \frac{\text{OPP}}{\text{HYP}} \qquad \csc A = \frac{\text{HYP}}{\text{OPP}}$$

$$\cos A = \frac{\text{ADJ}}{\text{HYP}} \qquad \sec A = \frac{\text{HYP}}{\text{ADJ}}$$

$$\tan A = \frac{\text{OPP}}{\text{ADJ}} \qquad \cot A = \frac{\text{ADJ}}{\text{OPP}}$$

Here are a couple of tips to help you remember these:

(1) Many students find it helpful to use the word SOHCAHTOA. You can think of the letters here as representing sin, opp, hyp, cos, adj, hyp, tan, opp, adj.

(2) The three trig functions on the right are the reciprocals of the three trig functions on the left. In other words, you get them by interchanging the numerator and denominator. It's pretty easy to remember that the reciprocal of tangent is cotangent. For the other two, just remember that the "s" goes with the "c" and the "c" goes with the "s." In other words, the reciprocal of sine is cosecant, and the reciprocal of cosine is secant.

To make sure you understand this, compute all six trig functions for each of the angles (except the right angle) in the triangle given in problem 30. Please try this yourself before looking at the answers below.

$\sin P = \frac{12}{13}$ $\quad \csc P = \frac{13}{12}$ $\quad \sin R = \frac{5}{13}$ $\quad \csc R = \frac{13}{5}$

$\cos P = \frac{5}{13}$ $\quad \sec P = \frac{13}{5}$ $\quad \cos R = \frac{12}{13}$ $\quad \sec R = \frac{13}{12}$

$\tan P = \frac{12}{5}$ $\quad \cot P = \frac{5}{12}$ $\quad \tan R = \frac{5}{12}$ $\quad \cot R = \frac{12}{5}$

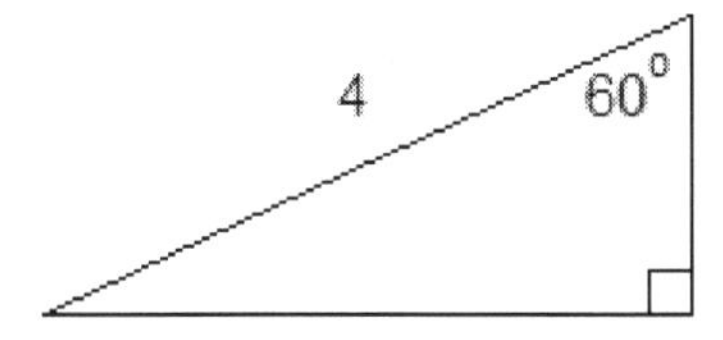


31. The figure above shows a right triangle whose hypotenuse is 4 feet long. How many feet long is the shorter leg of this triangle?

    (A) 2
    (B) 8
    (C) $2\sqrt{3}$
    (D) $\frac{2\sqrt{3}}{3}$
    (E) $\frac{8\sqrt{3}}{3}$

* The shorter leg of the triangle is adjacent to the 60° angle. So we will use cosine. We have $\cos 60° = \frac{\text{ADJ}}{\text{HYP}} = \frac{\text{ADJ}}{4}$. So ADJ = $4\cos 60° = 4(\frac{1}{2}) = 2$, choice (A)

**Remarks:** (1) If you do not see why we have $\cos 60° = \frac{\text{ADJ}}{\text{HYP}}$, review the basic trigonometry given after the solution to problem 30.

(2) To get from $\cos 60° = \frac{\text{ADJ}}{4}$ to ADJ = 4cos 60°, we simply multiply each side of the first equation by 4.

For those of you that like to cross multiply, the original equation can first be rewritten as $\frac{\cos 60°}{1} = \frac{\text{ADJ}}{4}$.

(3) There are several ways to compute cos 60°. The easiest is to simply put it into your calculator. The output will be .5.

(4) Make sure that your calculator is in degree mode before using your calculator. Otherwise you will get an incorrect answer.

If you are using a TI-84 (or equivalent) calculator press MODE and on the third line make sure that DEGREE is highlighted. If it is not, scroll down and select it. If possible do not alter this setting until you are finished taking your test.

(5) For the SAT Math Subject Test, it is worth knowing the following two special triangles:

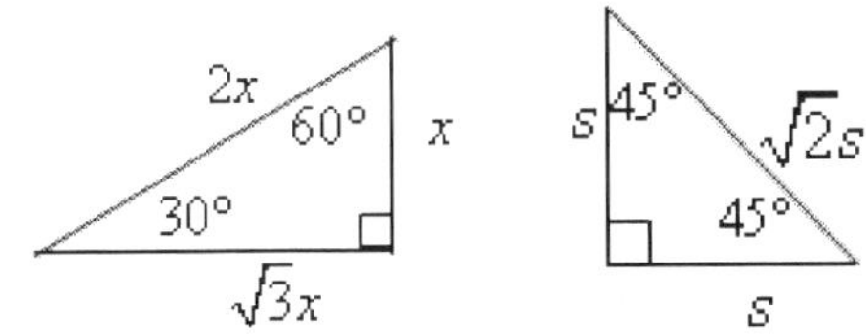


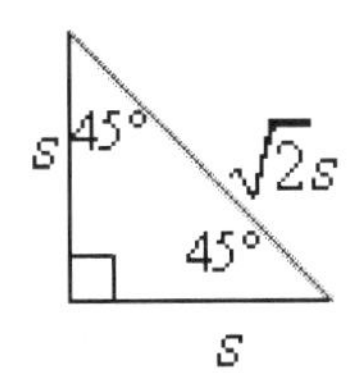


See the end of the solution to problem 23 for details. For this problem, we are using the 30, 60, 90 right triangle and $x = 3$.

For this problem, we have $\cos 60° = \frac{\text{ADJ}}{\text{HYP}} = \frac{x}{2x} = \frac{1}{2}$.

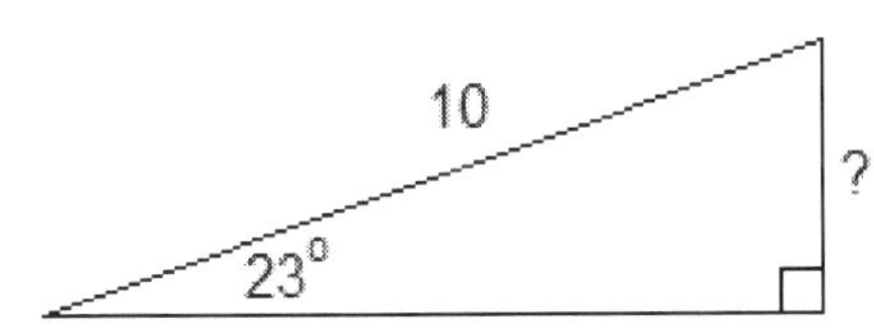


32. As shown above, a 10-foot ramp forms an angle of 23° with the ground, which is horizontal. Which of the following is an expression for the vertical rise, in feet, of the ramp?

(A) 10 cos 23°
(B) 10 sin 23°
(C) 10 tan 23°
(D) 10 cot 23°
(E) 10 sec 23°

* We have $\sin 23° = \frac{\text{OPP}}{\text{HYP}} = \frac{\text{OPP}}{10}$. So OPP = 10sin 23°, choice **G.**

**Remarks:** (1) If you do not see why we have $\sin 23° = \frac{\text{OPP}}{\text{HYP}}$, review the basic trigonometry given after the solution to problem 30.

(2) To get from $\sin 23° = \frac{\text{OPP}}{10}$ to OPP = 10sin 23°, we simply multiply each side of the first equation by 10.

For those of you that like to cross multiply, the original equation can first be rewritten as $\frac{\sin 23°}{1} = \frac{\text{OPP}}{10}$.

# LEVEL 2: NUMBER THEORY

33. A truck traveled 400 kilometers at an average speed of 60 kilometers per hour. Which of the following is the closest approximation to the amount of time that could be saved on this 400 kilometer trip if the average speed had increased 10 percent?

    (A) 15 minutes
    (B) 25 minutes
    (C) 35 minutes
    (D) 45 minutes
    (E) 1 hour

* **Solution using the formula $d = rt$:** We are given a distance of 400 and a rate of 60. So the total time is $\frac{400}{60} \approx 6.67$ hours.

If we increase the average speed by 10 percent, then the new rate is 1.1(60) = 66. So the new total time is $\frac{400}{66} \approx 6.06$ hours.

So the amount of time saved is approximately 6.67 – 6.06 = .61 hours.

We can change this to minutes by multiplying by 60. So the amount of time saved, in minutes, is approximately .61(60) = 36.6. The best approximation given in the answer choices is 35 minutes, choice (C).

**Notes:** (1) To change a percent to a decimal, divide by 100, or equivalently move the decimal point two places to the left (adding zeros if necessary). Note that the number 10 has an "invisible" decimal point after the 0 (so that 10 = 10.). Moving the decimal to the left two places gives us .10 = .1.

(2) To increase 60 by 10% we need to add 10% of 60 to the number 60 itself. There are two ways to do this.

**Method 1:** 10% of 60 = .1(60) = 6. So we get 60 + 6 = 66.

**Method 2:** 110% of 60 = 1.1(60) = 66.

34. The number line graph above is the graph of which of the following inequalities?

    (A) $-3 \leq x$ and $5 \leq x$
    (B) $-3 \leq x$ and $5 \geq x$
    (C) $-3 \leq x$ or $5 \leq x$
    (D) $-3 \geq x$ or $5 \leq x$
    (E) $-3 \geq x$ or $5 \geq x$

* The shaded part on the left represents all $x$-values less than or equal to $-3$, or equivalently, $x \leq -3$. This can also be written $-3 \geq x$.

The shaded part on the right represents all $x$-values greater than or equal to 5, or equivalently $x \geq 5$. This can also be written $5 \leq x$.

So the answer must be choice (D).

**Notes:** (1) The inequalities $x \geq a$ and $a \leq x$ are the same. Similarly, the inequalities $x \leq b$ and $b \geq x$ are the same.

(2) The word "or" indicates that $x$ can satisfy either of the two inequalities to be on the graph. This is why there are two pieces of the graph shaded.

(3) The word "and" indicates that both conditions need to be satisfied simultaneously. For example "$x = 3$ and $x = 4$" has no solutions since there are no $x$-values that are equal to both 3 and 4 at the same time. Similarly, "$-3 \geq x$ and $5 \leq x$" also has no solutions, and so it's graph would be empty (nothing would be shaded).

35. If $i = \sqrt{-1}$ and $((i^3)^5)^x = 1$, then the least positive integer value of $x$ is

    (A) 1
    (B) 2
    (C) 3
    (D) 4
    (E) 5

* **Solution by starting with choice A:** We start with choice A and type $((i^3)^5)^1$ into our calculator to get $-i$. So we can eliminate choice A.

Let's try choice B next to get $((i^3)^5)^2 = -1$.

Choice C gives us $((i^3)^5)^3 = i$.

Choice D gives $((i^3)^5)^4 = 1$. So the answer is choice (D).

**Notes:** (1) When plugging in answer choices we would usually start with choice C. An exception is when the word "least" appears in the problem. In this case we start with the smallest answer choice. this is why we start with choice A in this problem.

(2) Powers of $i$ can be computed quickly in your calculator, but be careful. Your calculator may sometimes "disguise" the number 0 with a tiny number in scientific notation. For example, let's compute $i^{73}$ using our calculator. When we type $i$ ^ 73 ENTER into our TI-84, we get an output of $-2.3\text{E}{-12} + i$. The expression $-2.3\text{E}{-12}$ represents a tiny number in scientific notation which is essentially 0. So this should be read as $0 + i = i$.

* **Algebraic solution:** $((i^3)^5)^x = (i^{15})^x = i^{15x}$. We are looking for the least positive integer value of $x$ so that $15x$ is a multiple of 4. The least such $x$ is $x = 4$, choice (D).

**Notes:** (1) See the table at the end of problem 4 for the law of exponents used here.

(2) Integer powers of $i$ are cyclical. Starting with $i^0 = 1$, we have

| | | | |
|---|---|---|---|
| $i^0 = 1$ | $i^1 = i$ | $i^2 = -1$ | $i^3 = -i$ |
| $i^4 = 1$ | $i^5 = i$ | $i^6 = -1$ | $i^7 = -i$ |
| $i^8 = 1$ | $i^9 = i$ | $i^{10} = -1$ | $i^{11} = -i$ |

...

In particular, when we raise $i$ to a nonnegative integer, there are only four possible answers: 1, $i$, $-1$, or $-i$

To decide which of these values is correct, we simply find the remainder upon dividing the exponent by 4.

For this problem, when we divide 15 by 4, the remainder is 3. Therefore $i^{15} = i^3 = -i$.

When we divide 30 by 4, the remainder is 2. So $i^{30} = i^2 = -1$.

When we divide 45 by 4, the reaminder is 1. So $i^{45} = i^1 = i$.

Finally, when we divide 60 by 4, the remainder is 0. So $i^{60} = i^0 = 1$.

36. For integers $a$ and $b$, if $a \odot b$ is the smallest integer divisible by both $a$ and $b$, then $(10 \odot 15) \odot 16 =$

    (A) 30
    (B) 240
    (C) 480
    (D) 1200
    (E) 2400

* $a \odot b$ is the least common multiple (or lcm) of $a$ and $b$. The lcm of 10 and 15 is 30, and the lcm of 30 and 16 is 240. Therefore

$$(10 \odot 15) \odot 16 = 30 \odot 16 = 240.$$

So the answer is choice (B).

**Computing the lcm:** Let's review two methods for computing the lcm of a set of numbers.

***Calculator method***

We use the **lcm** feature on our graphing calculator (found under NUM after pressing the MATH button). Our calculator can handle two numbers at a time. So compute **lcm**(10,15) = 30, and then **lcm**(30,16) = 240.

***Prime factorization method***

Step 1: Find the prime factorization of each integer in the set.

$$10 = 2 \cdot 5$$
$$15 = 3 \cdot 5$$
$$16 = 2^4$$

Step 2: Choose the highest power of each prime that appears in any of the factorizations and multiply them together:

$$2^4 \cdot 3 \cdot 5 = 240$$

Here is a quick lesson in **prime factorization:**

**The Fundamental Theorem of Arithmetic:** Every integer greater than 1 can be written "uniquely" as a product of primes.

The word "uniquely" is written in quotes because prime factorizations are only unique if we agree to write the primes in increasing order.

For example, 30 can be written as $2\cdot3\cdot5$ or as $3\cdot2\cdot5$ (as well as several other ways). But these factorizations are the same except that we changed the order of the factors.

To make things simple we always agree to use the **canonical representation**. The word "canonical" is just a fancy name for "natural," and the most natural way to write a prime factorization is in increasing order of primes. So the canonical representation of 30 is $2 \cdot 3 \cdot 5$.

As another example, the canonical representation of 100 is $2\cdot2\cdot5\cdot5$. We can tidy this up a bit by rewriting $2\cdot2$ as $2^2$ and $5\cdot5$ as $5^2$. So the canonical representation of 100 is $2^2\cdot5^2$.

If you are new to factoring, you may find it helpful to draw a factor tree. For example here is a factor tree for 100:

```
100
↙↘
[2]  50
     ↙↘
    [2]  25
         ↙↘
        [5] [5]
```

To draw this tree we started by writing 100 as the product $2 \cdot 50$. We put a box around 2 because 2 is prime, and does not need to be factored anymore. We then proceeded to factor 50 as $2 \cdot 25$. We put a box around 2, again because 2 is prime. Finally, we factor 25 as $5 \cdot 5$. We put a box around each 5 because 5 is prime. We now see that we are done, and the prime factorization can be found by multiplying all of the boxed numbers together. Remember that we will usually want the canonical representation, so write the final product in increasing order of primes.

By the Fundamental Theorem of Arithmetic above it does not matter how we factor the number – we will always get the same canonical form. For example, here is a different factor tree for 100:

100
↙↘
4 25
↙↘ ↙↘
[2] [2] [5] [5]

# LEVEL 2: ALGEBRA AND FUNCTIONS

37. If $a + a + a + a = b + b$, then $a - 2b =$

    (A) $-4a$
    (B) $-3a$
    (C) $-\frac{a}{2}$
    (D) $-\frac{a}{3}$
    (E) $-\frac{a}{4}$

* **Algebraic solution:** $a + a + a + a = 4a$ and $b + b = 2b$. So we have $4a = 2b$.We subtract $2b$ from each side to get $4a - 2b = 0$. Since we want $a - 2b$, we rewrite $4a$ as $a + 3a$. So we have $a - 2b + 3a = 0$. Subtracting $3a$ from each side gives $a - 2b = -3a$, choice (B).

**Solution by picking numbers:** Let's choose a value for $a$, say $a = 2$. Then $2 + 2 + 2 + 2 = b + b$. So $8 = 2b$, and therefore $b = 4$. It follows that $a - 2b = 2 - 8 = \mathbf{-6}$. Put a nice big dark circle around $\mathbf{-6}$ so you can find it easier later. We now substitute $a = 2$ into each answer choice:

(A) $-4a = -4(2) = -8$
(B) $-3a = -3(2) = -6$
(C) $-\frac{a}{2} = -\frac{2}{2} = -1$
(D) $-\frac{a}{3} = -\frac{2}{3}$
(E) $-\frac{a}{4} = -\frac{2}{4} = -\frac{1}{2}$

Since A, C, D, and E each came out incorrect, the answer is choice (B).

**Important note:** B is **not** the correct answer simply because it is equal to $-6$. It is correct because all four of the other choices are **not** $-6$. **You absolutely must check all five choices!**

38. If $f(x) = \frac{3x^2-5}{2x-1}$, what is $f(-0.01)$ ?

(A) 4.6
(B) 4.7
(C) 4.8
(D) 4.9
(E) 5.0

* $f(-0.01) = \frac{3(-0.01)^2-5}{2(-0.01)-1} \approx 4.9$, choice (D).

**Note:** This can be done in your calculator in one step:

(3*-.01 ^ 2 – 5) / (2*-.01 – 1) ENTER

39. For which nonnegative value of $a$ is the expression $\frac{1}{16-a^2}$ undefined?

(A) 0
(B) 4
(C) 16
(D) 32
(E) 64

* **Solution by plugging in answer choices:** We want to find a nonnegative value for $a$ that makes the denominator of the fraction zero. Normally we would start with choice C, but in this case it's pretty easy to see that choice B will work. Indeed, $16 - 4^2 = 16 - 16 = 0$. So the answer is choice (B).

**Algebraic solution:** The expression is undefined when the denominator is zero. So we need to solve the equation $16 - a^2 = 0$. Factoring the left hand side gives the equation $(4 - a)(4 + a) = 0$. So $4 - a = 0$ or $4 + a = 0$. Therefore we have $a = 4$ or $a = -4$. Since the question is asking for the nonnegative value of $a$, we choose $a = 4$, choice (B).

**Notes:** (1) The given expression is a **rational function**. A rational function is a quotient of polynomials (one polynomial divided by another polynomial). A rational function is undefined when the denominator is zero.

(2) The expression $16 - a^2$ is the **difference of two squares**. In general, the difference of two squares $x^2 - y^2$ factors as $(x - y)(x + y)$.

(3) We can also solve the equation $16 - a^2 = 0$ by adding $a^2$ to each side of the equation to get $16 = a^2$, and then using the **square root property** to get $\pm 4 = a$.

Note that the **square root property** says that if $x^2 = k^2$, then $x = \pm k$. This is different from taking a square root since it leads to two solutions.

40. What are all values of $x$ for which $|x + 3| \leq 4$?

    (A) $x \leq 1$
    (B) $x \leq -7$ or $x \geq 1$
    (C) $-7 \leq x \leq 1$
    (D) $-1 \leq x \leq 7$
    (E) $x \geq 1$

* **Algebraic solution:** The given absolute value inequality is equivalent to $-4 \leq x + 3 \leq 4$. We subtract 3 as follows:

$$\begin{array}{rcccl} -4 & \leq & x + 3 & \leq & 4 \\ \underline{-3} & & \underline{-3} & & \underline{-3} \\ -7 & \leq & x & \leq & 10 \end{array}$$

This is choice (C).

**An advanced lesson on absolute value and distance:**

**Geometrically**, $|x - y|$ is the distance between $x$ and $y$. In particular, $|x - y| = |y - x|$ .

**Examples:** $|5 - 3| = |3 - 5| = 2$ because the distance between 3 and 5 is 2.

If $|x - 3| = 7$, then the distance between $x$ and 3 is 7. So there are two possible values for $x$. They are $3 + 7 = 10$, and $3 - 7 = -4$. See the figure below for clarification.

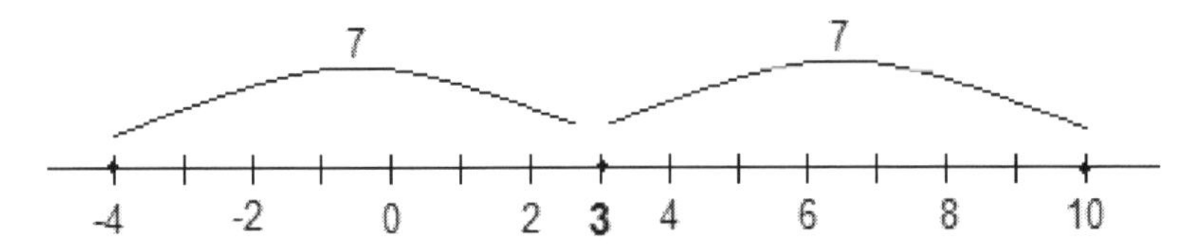


If $|x-3|<7$, then the distance between $x$ and 3 is less than 7. If you look at the above figure you should be able to see that this is all $x$ satisfying $-4<x<10$.

If $|x-3|>7$, then the distance between $x$ and 3 is greater than 7. If you look at the above figure you should be able to see that this is all $x$ satisfying $x<-4$ or $x>10$

**Algebraically**, we have the following. For $c>0$,

$$|x|=c \text{ is equivalent to } x=c \text{ or } x=-c$$

$$|x|<c \text{ is equivalent to } -c<x<c$$

$$|x|>c \text{ is equivalent to } x<-c \text{ or } x>c.$$

Let's look at the same examples as before algebraically.

**Examples:** If $|x-3|=7$, then $x-3=7$ or $x-3=-7$. So $x=10$ or $x=-4$.

If $|x-3|<7$, then $-7<x-3<7$. So $-4<x<10$.

If $|x-3|>7$, then $x-3<-7$ or $x-3>7$ . So $x<-4$ or $x>10$.

41. If $77^k=7^3\cdot 11^3$, what is the value of $k$ ?

    (A) 1
    (B) 3
    (C) 6
    (D) 9
    (E) 27

* **Algebraic solution:** $77^k=(7\cdot 11)^k=7^k\cdot 11^k$. So clearly $k$ = 3, choice (B).

**Note:** For the law of exponents used here see the table at the end of the solution to problem 4.

**Solution by starting with choice C:** We compute $7^3\cdot 11^3$ in our calculator and get 456,533.

We now start with choice C and guess that $k = 6$. When we put $77^6$ in our calculator we get a number that is much too large. So let's try choice B next and guess that $k = 3$. We see that $77^3 = 456{,}533$.

So the answer is choice (B).

42. For all $x \neq 0$, $\frac{1}{(\frac{5}{x^4})} =$

(A) $5x^4$

(B) $\frac{1}{5x^4}$

(C) $\frac{5}{x^4}$

(D) $\frac{x^4}{16}$

(E) $\frac{x^4}{5}$

* **Algebraic solution:** $\frac{1}{(\frac{5}{x^4})} = 1 \div \frac{5}{x^4} = 1 \cdot \frac{x^4}{5} = \frac{x^4}{5}$, choice (E).

**Remark:** This problem can also be solved by picking numbers. I leave the details of this method to the reader.

43. If $a$ is directly proportional to $b$ and $\frac{a}{b} = 3$, then what is the value of $b$ when $a = \frac{3}{2}$?

(A) $\frac{1}{4}$

(B) $\frac{1}{3}$

(C) $\frac{1}{2}$

(D) 1

(E) 2

* Since $a$ is directly proportional to $b$, $a = kb$ for some constant $k$, or equivalently, $\frac{a}{b} = k$. We are given that $\frac{a}{b} = 3$, so that $k = 3$. So we have $a = 3b$. When $a = \frac{3}{2}$, we have $\frac{3}{2} = 3b$. So $b = \frac{1}{2}$, choice (C).

Here is a quick lesson in **direct variation:**

The following are all equivalent ways of saying the same thing:

(1) $y$ varies directly as $x$
(2) $y$ is directly proportional to $x$
(3) $y = kx$ for some constant $k$
(4) $\frac{y}{x}$ is constant
(5) the graph of $y = f(x)$ is a nonvertical line through the origin.

For example, in the equation $y = 5x$, $y$ varies directly as $x$. Here is a partial table of values for this equation.

| $x$ | 1 | 2 | 3 | 4 |
|---|---|---|---|---|
| $y$ | 5 | 10 | 15 | 20 |

Note that we can tell that this table represents a direct relationship between $x$ and $y$ because $\frac{5}{1} = \frac{10}{2} = \frac{15}{3} = \frac{20}{4}$. Here the **constant of variation** is 5.

Here is a graph of the equation.

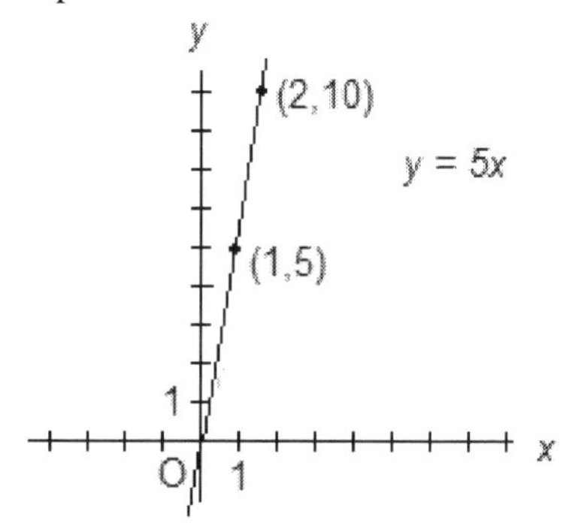


Note that we can tell that this graph represents a direct relationship between $x$ and $y$ because it is a nonvertical line through the origin. The constant of variation is the slope of the line, in this case $m = 5$.

44. If $y$ varies inversely as $x$ and $y = 8$ when $x = 3$, then what is $y$ when $x = 6$?

    (A) 1
    (B) 2
    (C) 3
    (D) 4
    (E) 6

**Solution 1:** Since $y$ varies inversely as $x$, $y = \frac{k}{x}$ for some constant $k$. We are given that $y = 8$ when $x = 3$, so that $8 = \frac{k}{3}$, or $k = 24$. Thus, $y = \frac{24}{x}$. When $x = 6$, we have $y = \frac{24}{6} = 4$, choice (D).

**Solution 2:** Since y varies inversely as $x$, $xy$ is a constant. So we get the following equation: $(3)(8) = 6y$ So $24 = 6y$, and $y = \frac{24}{6} = 4$, choice (D).

* **Quick computation:** $\frac{(8)(3)}{6} = 4$, choice (D).

Here is a quick lesson in **inverse variation:**

The following are all equivalent ways of saying the same thing:

(1) $y$ varies inversely as $x$
(2) $y$ is inversely proportional to $x$
(3) $y = \frac{k}{x}$ for some constant $k$
(4) $xy$ is constant

The following is a consequence of (1), (2) (3) or (4).

(5) The graph of $y = f(x)$ is a hyperbola.

**Note:** (5) is not equivalent to (1), (2), (3) or (4).

For example, in the equation $y = \frac{12}{x}$, $y$ varies inversely as $x$. Here is a partial table of values for this equation.

| $x$ | 1 | 2 | 3 | 4 |
|---|---|---|---|---|
| $y$ | 12 | 6 | 4 | 3 |

Note that we can tell that this table represents an inverse relationship between $x$ and $y$ because $(1)(12) = (2)(6) = (3)(4) = (4)(3) = 12$. Here the **constant of variation** is 12.

Here is a graph of the equation. On the left you can see the full graph. On the right we have a close-up in the first quadrant.

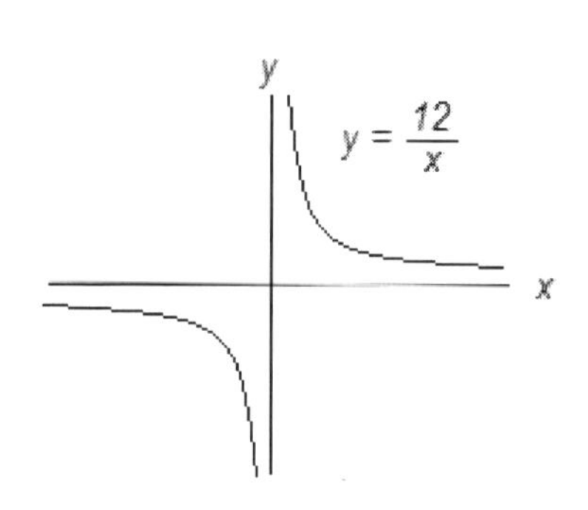


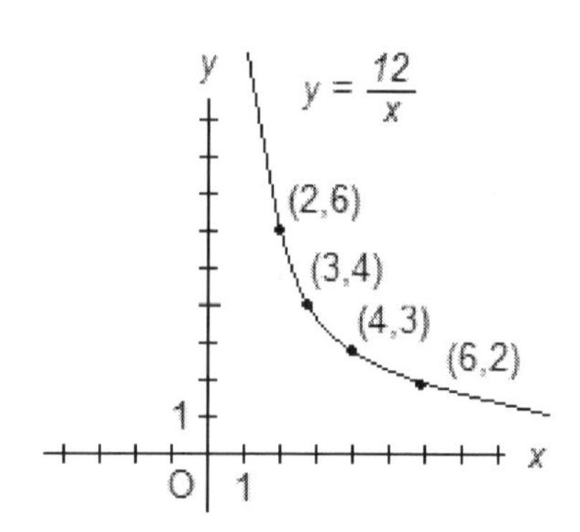


45. The operation & is defined as $r \,\&\, s = \frac{s^2 - r^2}{r + s}$ where $r$ and $s$ are real numbers and $r \neq -s$. What is the value of $(-3) \,\&\, (-4)$ ?

    (A) 2
    (B) 1
    (C) 0
    (D) −1
    (E) −2

* $(-3) \,\&\, (-4) = \frac{(-4)^2 - (-3)^2}{(-3)+(-4)} = \frac{16-9}{-3-4} = \frac{7}{-7} = -1$, choice (D).

46. In the real numbers, what is the solution of the equation $4^{x+2} = 8^{2x-1}$ ?

    (A) $-\frac{7}{4}$

    (B) $-\frac{1}{4}$

    (C) $\frac{3}{4}$

    (D) $\frac{5}{4}$

    (E) $\frac{7}{4}$

* **Algebraic solution:** The numbers 4 and 8 have a common base of 2. In fact, $4 = 2^2$ and $8 = 2^3$. So we have $4^{x+2} = (2^2)^{x+2} = 2^{2x+4}$ and we have $8^{2x-1} = (2^3)^{2x-1} = 2^{6x-3}$. Thus, $2^{2x+4} = 2^{6x-3}$. So $2x + 4 = 6x - 3$. We subtract $2x$ from each side of this equation to get $4 = 4x - 3$. We now add 3 to each side of this last equation to get $7 = 4x$. Finally we divide each side of this equation by 4 to get $\frac{7}{4} = x$, choice (E).

**Notes:** (1) For a review of the laws of exponents used here see the end of the solution to problem 4.

(2) This problem can also be solved by plugging in (start with choice C). We leave it to the reader to solve the problem this way. Make sure to use your calculator.

# LEVEL 2: GEOMETRY

47. Triangle $PQR$ is similar to triangle $STU$. The length of side $QR$ is 4.3 inches, the length of corresponding side $TU$ is 5.2 inches, and the perimeter of triangle $PQR$ is 7.1 inches. What is the perimeter of triangle $STU$?

    (A) 3.1 in
    (B) 4.3 in
    (C) 5.8 in
    (D) 5.9 in
    (E) 8.6 in

* **Solution by setting up a ratio:** We begin by identifying the side lengths and perimeters involved in the problem.

| | | |
|---|---|---|
| length | 4.3 | 5.2 |
| perimeter | 7.1 | $P$ |

Now draw in the division symbols and equal sign, cross multiply and divide the corresponding ratio to find the unknown quantity $P$.

$$\frac{4.3}{7.1} = \frac{5.2}{P}$$

$$4.3P = 36.92$$

$$P \approx 8.6$$

So the answer is choice (E).

**Notes:** (1) Two triangles are **similar** if their angles are congruent.

(2) Similar triangles **do not** have to be the same size, but **corresponding sides of similar triangles are in proportion**. So for example,

$$\frac{QR}{TU} = \frac{PQ}{ST}$$

(This particular proportion is not needed in this problem).

(3) Since all sides lengths of similar triangles are in proportion, the ratio of a side length to the perimeter is the same in similar triangles as well (as long as corresponding sides are used).

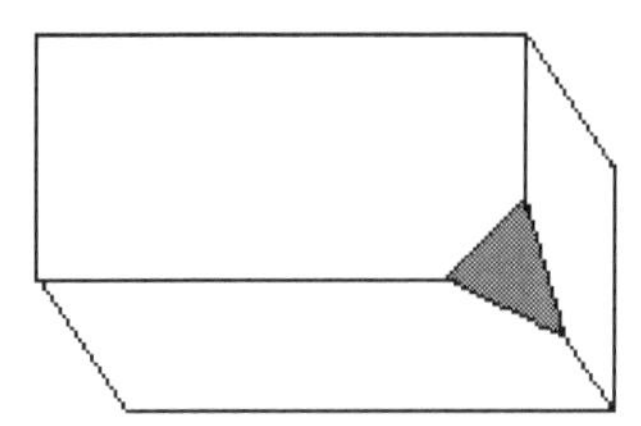

48. A tetrahedron was cut from the corner of the rectangular solid shown, with each of its vertices less than halfway to the midpoint of each edge of the solid. If tetrahedrons of the same size are cut from the remaining seven corners of the rectangular solid, how many edges will the resulting solid have?

    (A) 12
    (B) 14
    (C) 24
    (D) 36
    (E) 42

* The original solid had 8 vertices and 12 edges. Each tetrahedron adds 3 more edges, and there are 8 tetrahedrons (one for each vertex). So we get a total of $12 + 8(3) = 36$ edges, choice (D).

49. What is the $x$-coordinate of the point at which the line whose equation is $5x - 4y + 3 = 0$ crosses the $x$-axis?

    (A) $-\frac{5}{3}$
    (B) $-\frac{3}{5}$
    (C) $\frac{3}{5}$
    (D) $\frac{3}{4}$
    (E) $\frac{5}{3}$

* The line crosses the $x$-axis when $y = 0$. So $5x - 4(0) + 3 = 0$, or equivalently $5x + 3 = 0$. We subtract 3 from each side of this equation to get $5x = -3$. Finally, we divide by 5 to get $x = -\frac{3}{5}$, choice (B).

50. If lines $m$ and $n$ are parallel and are intersected by transversal $L$, what is the sum of the measures of the exterior angles on the same side of line $L$ ?

    (A) 45°
    (B) 90°
    (C) 180°
    (D) 270°
    (E) 360°

* The exterior angles on the same side of a transversal are not congruent, and therefore they are supplementary. So the sum of their measures is 180°, choice (C).

Let's look at a picture.

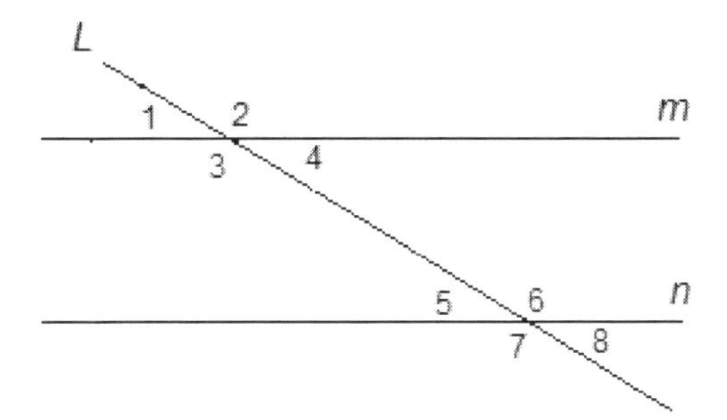


In the following picture, angles 1, 4, 5, and 8 are congruent. Also angles 2, 3, 6, and 7 are congruent. Any two angles that are not congruent are supplementary. In particular, the exterior angles 2 and 8 are supplementary. Note also that angles 2 and 8 are on the same side of the transversal.

Note that angles 1 and 7 are also exterior angles on the same side of the transversal. They are supplementary as well, of course.

51. The consecutive vertices of a certain isosceles trapezoid are $P$, $Q$, $R$, and $S$ where $\overline{PQ} \parallel \overline{SR}$. Which of the following are NOT congruent?

    (A) $\angle P$ and $\angle Q$
    (B) $\angle R$ and $\angle S$
    (C) $\overline{PS}$ and $\overline{QR}$
    (D) $\overline{PQ}$ and $\overline{SR}$
    (E) $\overline{PR}$ and $\overline{QS}$

* **Solution by drawing a picture:** Let's draw a picture

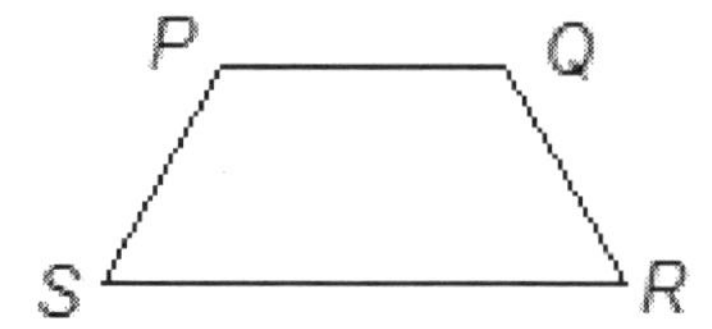


From the picture we see that $\overline{PQ}$ and $\overline{SR}$ are not congruent, choice (D).

**Note:** The symbol || means "is parallel to."

52. Which of the following is an equation of a line that is parallel to the line with equation $5x - 3y = 2$ ?

    (A) $y = \frac{5}{3}x + 1$

    (B) $y = -\frac{5}{3}x - 1$

    (C) $y = \frac{3}{5}x + 2$

    (D) $y = -\frac{3}{5}x - 3$

    (E) $y = 5x$

* We first find the slope of the given line by putting it into slope-intercept form. In other words we solve for $y$.

$$5x - 3y = 2$$
$$-3y = -5x + 2$$
$$y = \frac{5}{3}x - \frac{2}{3}$$

From this last equation we see that the given line has a slope of $\frac{5}{3}$. So the slope of a line parallel to this one is also $\frac{5}{3}$. So the answer is choice (A).

**Notes:** (1) See the end of problem 18 for more information on slope and slope-intercept form.

(2) Parallel lines have the same slope.

(3) The slope of a line in the **general form** $ax + by = c$ is $-\frac{a}{b}$. If you choose to memorize this fact, you can find the slope of the line given in this problem quickly without first rewriting the equation in slope-intercept form.

In this question $a = 5$ and $b = -3$. So the slope of the line with equation $5x - 3y = 2$ is $-\frac{5}{-3} = \frac{5}{3}$.

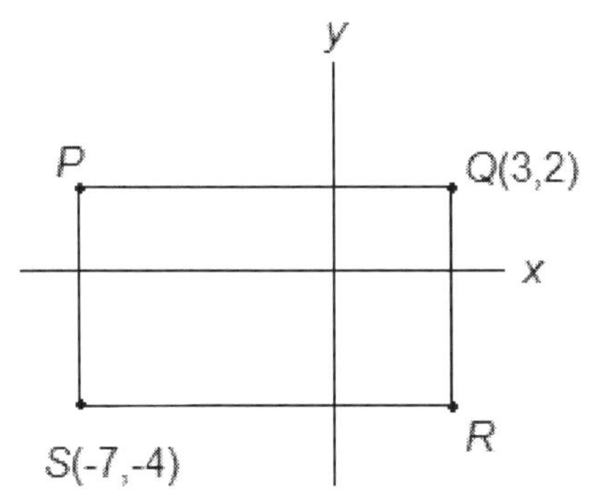


53. In the figure above, the sides of rectangle $PQRS$ are parallel to the axes. What is the distance between points $P$ and $R$ ?

    (A) 8.5
    (B) 10.2
    (C) 11.7
    (D) 12.3
    (E) 15.1

**Solution using the Pythagorean Theorem:** First note that point $P$ has the same $x$-coordinate as point $S$ and the same $y$-coordinate as point $Q$. So point $P$ has coordinates $(-7,2)$.

Similarly, point $R$ has the same $x$-coordinate as point $Q$ and the same $y$-coordinate as point $S$. So point $R$ has coordinates $(3,-4)$.

We now form the right triangle $PSR$, and note that $PS = 2 - (-4) = 6$ and $SR = 3 - (-7) = 10$.

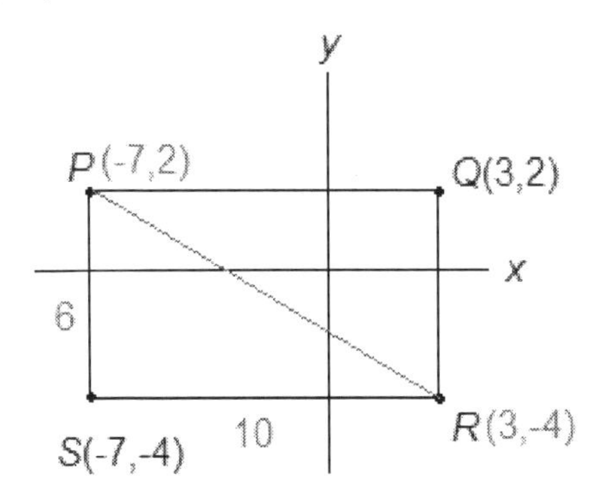


We can now use the Pythagorean Theorem to get

$$PR^2 = 6^2 + 10^2 = 36 + 100 = 136.$$

So $PR = \sqrt{136} \approx 11.7$, choice (C).

**Note:** The Pythagorean Theorem says that if a right triangle has legs of length $a$ and $b$, and a hypotenuse of length $c$, then $c^2 = a^2 + b^2$.

**Solution using the distance formula:** As in the previous solution we note that $P$ has coordinates $(-7,2)$ and $R$ has coordinates $(3,-4)$. We now use the distance formula to get

$$PR = \sqrt{(3-(-7))^2 + (-4-2)^2} = \sqrt{10^2 + (-6)^2} = \sqrt{136} \approx 11.7.$$

This is choice (C).

**Note:** The distance between the points $(s, t)$ and $(u, v)$ is

$$d = \sqrt{(u-s)^2 + (v-t)^2}$$

54. In the standard $(x, y)$ coordinate plane, point $M$ with coordinates $(3,7)$ is the midpoint of $\overline{PQ}$, and $P$ has coordinates $(1,9)$. What are the coordinates of $Q$ ?

    (A) $(5,5)$
    (B) $(-5,-5)$
    (C) $(-1,11)$
    (D) $(2,8)$
    (E) $(4,16)$

* **Solution by counting:** To get from point $P$ to point $M$ we need to move right 2 units and down 2 units. So to get from point $M$ to point $Q$ we must also move right 2 units and down 2 units. So the coordinates of point $Q$ are $(5,5)$, choice (A).

**Solution using the midpoint formula:** If point $P$ has coordinates $(x_1, y_1)$ and point $Q$ has coordinates $(x_2, y_2)$, then the midpoint $M$ has coordinates $(\frac{x_1+x_2}{2}, \frac{y_1+y_2}{2})$.

In this case we have $x_1 = 1$, $y_1 = 9$, $\frac{x_1+x_2}{2} = 3$, and $\frac{y_1+y_2}{2} = 7$.

Substituting $x_1 = 1$ into the third equation we have $\frac{1+x_2}{2} = 3$. We multiply by 2 to get $1 + x_2 = 6$. We subtract 1 to get $x_2 = 5$.

Substituting $y_1 = 9$ into the fourth equation we have $\frac{9+y_2}{2} = 7$. We multiply by 2 to get $9 + y_2 = 14$. We subtract 9 to get $y_2 = 5$.

So point $Q$ has coordinates (5,5), choice (A).

55. Cube $P$ is inscribed in sphere $O$, and sphere $O$ is inscribed in cube $Q$. Which of the following statements must be true?

    (A) The length of a long diagonal of $Q$ is equal to the length of a diameter of $O$.
    (B) The length of a long diagonal of $P$ is equal to the length of an edge of $Q$.
    (C) The length of an edge of $P$ is equal to the length of a diameter of $O$.
    (D) The length of an edge of $Q$ is equal to the length of a radius of $O$.
    (E) The length of a long diagonal of $P$ is equal to the length of a radius of $O$.

* A long diagonal of cube $P$ is a diameter of sphere $O$. Also, an edge of cube $Q$ has the same length as a diameter of sphere $O$, Therefore the length of a long diagonal of $P$ is equal to the length of an edge of $Q$, choice (B).

**Remark:** Below are pictures of cube $P$ inscribed in sphere $O$ and sphere $O$ inscribed in cube $Q$.

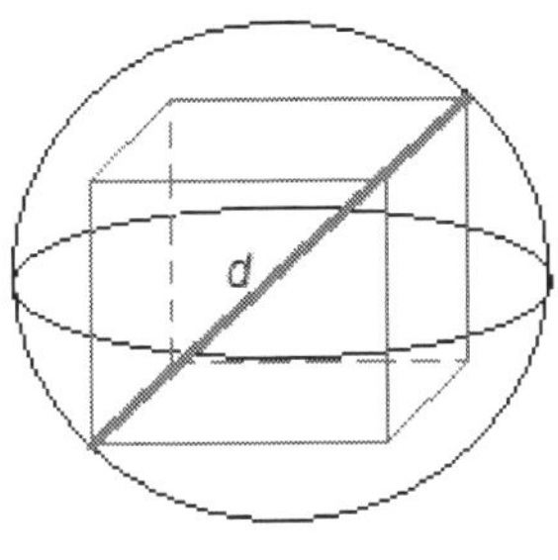


Cube $P$ inscribed in sphere O

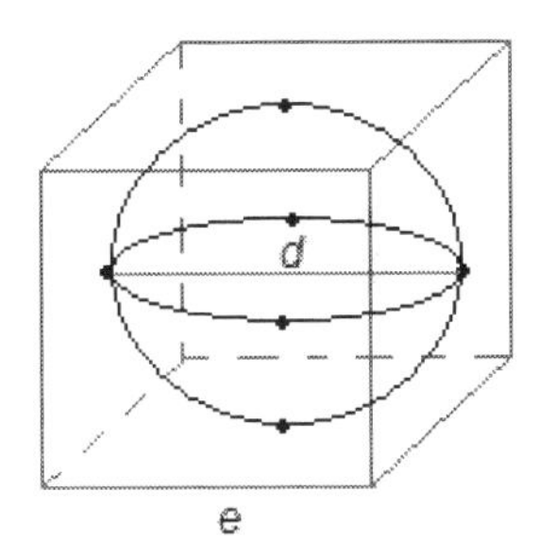


Sphere O inscribed in cube Q

Note that in the left picture that the long diagonal $d$ of cube $P$ is a diameter of sphere $O$. Also note that in the right picture that the diameter $d$ of sphere $O$ has the same length as edge $e$ of cube $Q$.

56. If $a > 1$, what is the slope of the line in the $xy$-plane that passes through the points $(a^2, a^4)$ and $(a^3, a^6)$?

    (A) $-a^3 + 6a^2$
    (B) $-a^3 + a^2$
    (C) $-a^3 - a^2$
    (D) $a^3 - a^2$
    (E) $a^3 + a^2$

**Solution by picking numbers:** Let's pick a number for $a$, say $a = 2$. So the two points are (4,16) and (8,64). The slope of the line passing through these two points is

$$m = \frac{64-16}{8-4} = \frac{48}{4} = \mathbf{12}$$

**Put a nice big, dark circle around the number 12.** We now plug $a = 2$ into each answer choice.

(A) -8 + 6(4) = 16
(B) -8 + 4 = -4
(C) -8 – 4 = -12
(D) 8 – 4 = 4
(E) 8 + 4 = 12

Since choices A, B, C, and D all came out incorrect, the answer is choice (E).

**Remark:** We could have also gotten the slope geometrically by plotting the two points, and noticing that to get from (4,16) to (8,64) we need to travel up 48 units and right 4 units. So the slope is

$$m = \frac{rise}{run} = \frac{48}{4} = \mathbf{12}.$$

Before we go on, try to solve this problem directly (without plugging in numbers).

* **Solution using the slope formula:** Using the slope formula we have

$$m = \frac{a^6-a^4}{a^3-a^2} = \frac{a^4(a^2-1)}{a^2(a-1)} = \frac{a^2(a+1)(a-1)}{a-1} = a^2(a+1) = a^3 + a^2.$$

This is choice (E).

# LEVEL 2: PROBABILITY AND STATISTICS

57. Five shapes – a triangle, a square, a trapezoid, a pentagon, and a hexagon – are arranged in a row from left to right. How many such arrangements are there where the square is in the middle?

    (A) 4
    (B) 5
    (C) 24
    (D) 25
    (E) 120

**Solution using the counting principle:** Consider 5 positions, and place the square in the middle position. There are now 4 positions left. So there are 4 positions to place the triangle. Once the triangle is placed, there are 3 positions to place the trapezoid. Next there are 2 positions to place the pentagon. Finally, there is 1 position left to place the hexagon. By the counting principle we get (4)(3)(2)(1) = 24 arrangements, choice (C).

**Note:** See problem 26 for more information on the counting principle

* **Solution using permutations:** The square is placed automatically, and there are 4 shapes left to arrange.. So there are ${}_4P_4 = 4! = 24$ arrangements, choice (C).

**Remarks:** (1) This is a permutation because we are arranging the shapes.

(2) We can compute ${}_4P_4$ very quickly on our calculator as follows: first type 4. Then under the Math menu scroll over to PRB and select nPr. Finally type 4 and press ENTER. You will get an answer of 24.

(3) The formula for $nPr$ is $\frac{n!}{(n-r)!}$. So ${}_4P_4 = \frac{4!}{0!} = 24$. (Note that this is included for completeness. You do not need to know this formula.)

58. If $A = \{1,2,3,4,5,6,7,8,9,10\}$ and $B = \{3,6,9,12,15\}$, what is the mean of $A \cup B$ ?

    (A) 5.33
    (B) 5.82
    (C) 5.96
    (D) 6.24
    (E) 6.83

* $A \cup B = \{1,2,3,4,5,6,7,8,9,10,12,15\}$. Therefore the mean of $A \cup B$ is $\frac{1+2+3+4+5+6+7+8+9+10+12+15}{12} = \frac{82}{12} \approx 6.83$, choice (E).

**Notes:** (1) $A \cup B$ is read "the **union** of $A$ and $B$." It is the set consisting of the elements that are in $A$ or $B$ or both.

(2) The **average (arithmetic mean)** of a list of numbers is the sum of the numbers in the list divided by the quantity of the numbers in the list.

$$\textbf{Average} = \frac{\textbf{Sum}}{\textbf{Number}}$$

59. In Bakerfield, 60% of the population own at least 1 cat. 20% of the cat owners in Bakerfield play the piano. If a resident of Bakerfield is selected at random, what is the probability that this person is a piano player that owns at least 1 cat?

(A) 0.12
(B) 0.15
(C) 0.33
(D) 0.37
(E) 0.80

* Let $E$ be the event "owns at least 1 cat," and let $F$ be the event "plays the piano." We are given $P(E) = .6$ and $P(F|E) = .2$. It follows that $P(E \cap F) = P(E) \cdot P(F|E) = (0.6)(0.2) = 0.12$, choice (A).

**Notes:** (1) To change a percent to a decimal, divide by 100, or equivalently move the decimal point two places to the left (adding zeros if necessary). Note that the number 60 has an "invisible" decimal point after the 0 (so that 60 = 60.). Moving the decimal to the left two places gives us .60 = .6.

(2) "60% of the population own at least 1 cat" is equivalent to "the probability that someone from the population owns a cat is .6." This was written above symbolically as $P(E) = .6$.

Similarly, "20% of the cat owners in Bakerfield play the piano" is equivalent to "the probability that someone from Bakerfield plays the piano **given** that this person owns a cat is .2." This was written above symbolically as $P(F|E) = .2$. Note that the symbol | is read "given," so that $P(F|E)$ is read "the probability of $F$ given $E$."

(3) $E \cap F$ is read "the **intersection** of $E$ and $F$. " It is the event consisting of the outcomes that are common to both $E$ and $F$. In this problem a member of $E \cap F$ is a person from Bakerfield that owns at least 1 cat **and** plays the piano.

(4) $P(F|E)$ is called a **conditional probability**. The conditional probability formula is $\boldsymbol{P(E \cap F) = P(E) \cdot P(F|E)}$.

60. How many committees of 5 people can be formed from a group of 10 people?

    (A) 5
    (B) 10
    (C) 50
    (D) 252
    (E) 30,240

* This is a **combination**. The answer is ${}_{10}C_5 = 252$, choice (D).

**Remarks:**

(1) This is a combination because it does not matter in what order we take the 5 people. We are simply putting the 5 people together into a group.

(2) We can compute ${}_{10}C_5$ very quickly on our calculator as follows: first type 10. Then under the Math menu scroll over to PRB and select nCr. Finally type 5 and press ENTER.

(3) The formula for ${}_nC_r$ is $\frac{n!}{r!(n-r)!}$. So ${}_{10}C_5 = \frac{10!}{5!5!} = 252$. (Note that this is included for completeness. You do not need to know this formula.)

# LEVEL 2: TRIGONOMETRY

61. In $\Delta CAT$, $\angle A$ is a right angle. Which of the following is equal to $\tan T$?

    (A) $\frac{CA}{CT}$
    (B) $\frac{CA}{AT}$
    (C) $\frac{CT}{CA}$
    (D) $\frac{CT}{AT}$
    (E) $\frac{AT}{CA}$

* **Solution by drawing a picture:** Let's draw a picture.

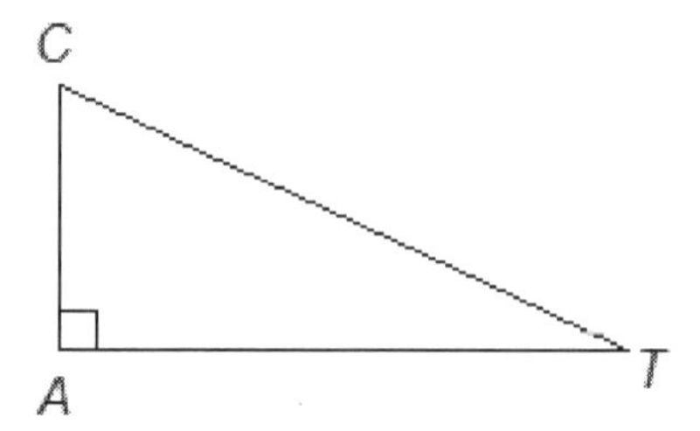


Now just note that $\tan T = \frac{\text{OPP}}{\text{ADJ}} = \frac{CA}{AT}$, choice (B).

**Remark:** If you do not see why we have $\tan T = \frac{\text{OPP}}{\text{ADJ}}$, review the basic trigonometry given after the solution to problem 30.

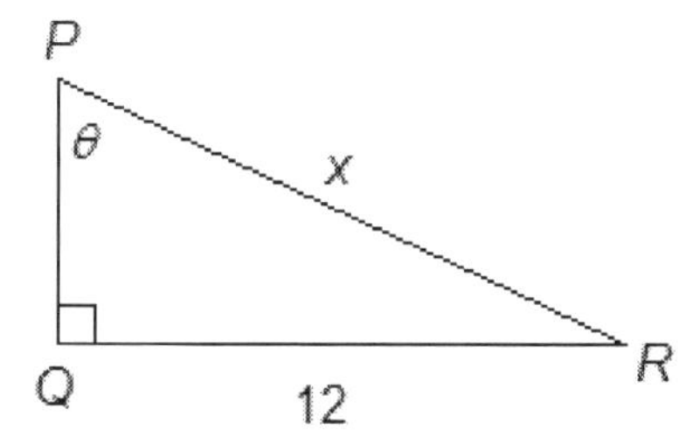


62. In the figure above $\theta = 65°$. What is the value of $x$ ?

 (A) 5.07
 (B) 5.60
 (C) 10.87
 (D) 13.24
 (E) 28.39

* $\sin\theta = \frac{\text{OPP}}{\text{HYP}} = \frac{12}{x}$. Multiplying each side of this equation by $x$ gives us $x\sin\theta = 12$. So $x = \frac{12}{\sin\theta} = \frac{12}{\sin 65°} \approx 13.24$, choice (D).

**Remarks:** (1) If you do not see why we have $\sin\theta = \frac{\text{OPP}}{\text{HYP}}$, review the basic trigonometry given after the solution to problem 30.

(2) If you prefer, you can think of the multiplication above as **cross multiplication** by first rewriting $\sin\theta$ as $\frac{\sin\theta}{1}$.

So we have $\frac{\sin\theta}{1} = \frac{12}{x}$. Cross multiplying yields $x\sin\theta = 1(12)$. This yields the same result as in the above solution.

63. In $\Delta ABC$ with right angle $C$, $BC = 5$ and $\cos B = 0.7$. What is the length of $AB$ ?

(A) 0.14
(B) 0.29
(C) 3.50
(D) 5.26
(E) 7.14

* **Solution by drawing a picture:** Let's draw a picture.

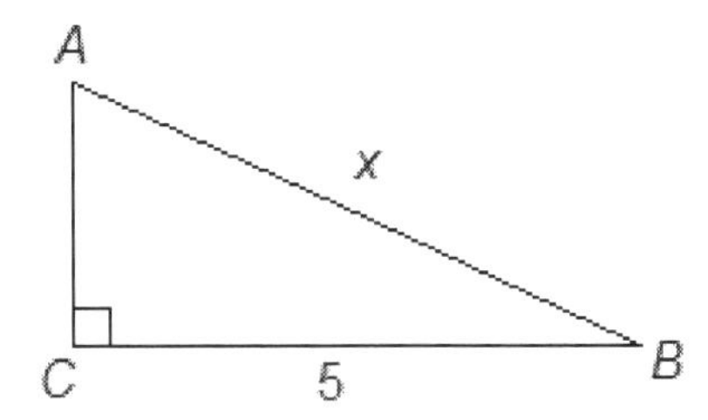


$\cos B = \frac{\text{ADJ}}{\text{HYP}}$. So we have $0.7 = \frac{5}{x}$. Multiplying each side of this equation by $x$ gives us $0.7x = 5$. So $x = \frac{5}{0.7} \approx 7.14$, choice (E).

64. For any acute angle with measure $A$, $\sin(90° - A) =$

(A) $\sin A$
(B) $\cos A$
(C) $\tan A$
(D) $\csc A$
(E) $\sec A$

**Solution by picking a number:** Let's let $A = 5°$. We use our calculator to get $\sin(90° - A) = \sin(85°) \approx$ **0.996**. Put a nice big, dark circle around **0.996** so you can find it easily later. We now substitute our value for $A$ into each answer choice.

(A) 0.087
(B) 0.996
(C) 0.087
(D) 11.474
(E) 1.004

Since A, C, D and E are incorrect we can eliminate them. Therefore the answer is choice (B).

* **Solution using an identity:** We use the following difference identity:

$$\mathbf{\sin(x - y) = \sin x \cos y - \cos x \sin y}$$

$$\sin(90° - A) = \sin 90° \cos A - \cos 90° \sin A$$

$\sin 90° = 1$ and $\cos 90° = 0$, so this last expression is $\cos A$, choice (B).

**Cofunction Identities:** You do not NEED to memorize these, but you can if you want:

$$\mathbf{\sin(90° - A) = \cos A} \qquad \mathbf{\csc(90° - A) = \sec A}$$

$$\mathbf{\cot(90° - A) = \tan A}$$

# LEVEL 3: NUMBER THEORY

65. If $k$ is a negative integer, for which of the following values of $n$ is $|n^4 - k|$ greatest?

    (A) −4
    (B) −1
    (C) 0
    (D) 2
    (E) 3

* **Solution by picking a number:** Let's choose a value for $k$, say $k = -2$. Then we are looking for the greatest value of

$$|n^4 - k| = |n^4 - (-2)| = |n^4 + 2|.$$

This will be large when $n^4$ is large, and so the answer is $-4$, choice (A).

**Notes:** (1) Since $n^4 + 2$ is always positive, $|n^4 + 2| = n^4 + 2$.

(2) $n^4$ is nonnegative, and so $n^4$ is largest when $|n|$ is largest. Of the answer choices in this problem, $-4$ has the largest absolute value.

(3) If you are unsure which of the five answer choices gives the largest value, you can also plug in all 5 answer choices.

66. If $\log_b x = k$, which of the following must be true?

(A) $b^x = k$
(B) $b^k = x$
(C) $x^b = k$
(D) $x^k = b$
(E) $k^b = x$

* **Solution by changing to exponential form:** We simply change the logarithmic equation to exponential form to get $b^k = x$, choice (B).

**Notes:** (1) The word "logarithm" just means "exponent."

(2) The equation $y = \log_b x$ can be read as "$y$ is the exponent when we rewrite $x$ with a base of $b$." In other words we are raising $b$ to the power $y$. So the equation can be written in exponential form as $x = b^y$.

(3) The base $b$ of a logarithm must satisfy $b > 0$ and $b \neq 1$.

**Alternate solution:** We exponentiate each side of the equation using base $b$ to get $b^{\log_b x} = b^k$. Since $b^{\log_b x} = x$, we get $x = b^k$. This is the same as choice (B).

**Note:** Recall that the functions $b^x$ and $\log_b x$ are inverses of each other. This means that $b^{\log_b x} = x$ and $\log_b b^x = x$ (we needed only the first equation for this solution).

67. In an arithmetic sequence, the third term is 7 and the eighth term is 27. What is the tenth term in the sequence?

(A) 35
(B) 34
(C) 33
(D) 32
(E) 31

* **Quick solution:** We can find the common difference of this arithmetic sequence with the computation

$$d = \frac{27-7}{8-3} = \frac{20}{5} = 4.$$

The ninth term is 27 + 4 = 31, the tenth term is 31 + 4 = 35, choice (A).

**Remarks:** (1) In an arithmetic sequence, you always add (or subtract) the same number to get from one term to the next. This can be done by moving forwards or backwards through the sequence.

(2) Questions about arithmetic sequences can easily be thought of as questions about lines and linear equations. We can identify terms of the sequence with points on a line where the $x$-coordinate is the term number and the $y$-coordinate is the term itself.

In the question above, since the third term of the sequence is 7, we can identify this term with the point (3,7). Since the eighth term of the sequence is 27, we can identify this with the point (8,27). Note that the common difference $d$ is just the slope of the line that passes through these two points, i.e. $d = \frac{27-7}{8-3} = 4$.

**Definition:** An **arithmetic sequence** is a sequence of numbers such that the difference $\boldsymbol{d}$ between consecutive terms is constant. The number $\boldsymbol{d}$ is called the **common difference** of the arithmetic sequence.

Example of an arithmetic sequence: –1, 3, 7, 11, 15, 19, 23, 27, 31, 35,…

In this example the common difference is $d = 3 - (-1) = 4$.

Note that this is the same arithmetic sequence given in this question.

**Arithmetic sequence formula:** $\boldsymbol{a_n = a_1 + (n-1)d}$

In the above formula, $a_n$ is the $n$th term of the sequence. For example, $a_1$ is the first term of the sequence.

**Note:** In the arithmetic sequence –1, 3, 7, 11, 15, 19, 23, 27, 31, 35,… we have that $a_1 = -1$ and $d = 4$. Therefore

$$a_n = -1 + (n-1)(4) = -1 + 4n - 4 = -5 + 4n.$$

It follows that $a_{10} = -5 + 4(10) = -5 + 40 = 35$, choice (A).

**Solution using the arithmetic sequence formula** Substituting 3 in for $n$ and 7 in for $a_n$ into the arithmetic sequence formula gives us $7 = a_1 + 2d$.

Similarly, substituting 8 in for $n$ and 27 in for $a_n$ into the arithmetic sequence formula gives us $27 = a_1 + 7d$.

So we solve the following system of equations to find $d$.

$$\begin{aligned} 27 &= a_1 + 7d \\ \underline{7} &\underline{= a_1 + 2d} \\ 20 &= \quad\;\; 5d \end{aligned}$$

The last equation comes from subtraction. We now divide each side of this last equation by 5 to get $d = 4$.

Finally, we add 4 to 27 twice to get $27 + 4(2) = 35$, choice (A).

**Remarks:** (1) We used the elimination method to find $d$ here. This is usually the quickest way to solve a system of linear equations on this test.

(2) Once we have that $d = 4$, we can substitute this into either of the original equations to find $a_1$. For example, we have $7 = a_1 + 2(4)$, so that $a_1 = 7 - 8 = -1$.

68. A deposit of $800 is made into an account that earns 2% interest compounded annually. If no additional deposits are made, how many years will it take until there is $990 in the account?

    (A) 9
    (B) 10
    (C) 11
    (D) 12
    (E) 13

* We use the formula $A = P(1 + r)^t$ for interest compounded annually. We are given that $P = 800$, $r = 0.02$, $A = 990$, and we want to find $t$. So we have $990 = 800(1.02)^t$. We can now proceed in 2 ways.

**Method 1 – Starting with choice C:** We start with choice (C) and substitute 11 in for $t$ to get $800(1.02)^{11} \approx 994.7$. So the answer is (C).

Note that $800(1.02)^{10} \approx 975$, and this is too small.

**Method 2 – Algebraic solution:** We divide each side of the equation by 800 to get $1.2375 = (1.02)^t$. We then take the natural logarithm of each side to get $\ln 1.2375 = \ln(1.02)^t$. We can now use a basic law of logarithms to bring the $t$ out in front of 1.02 (see the last row of the table below). We get $\ln 1.2375 = t \ln 1.02$.

Finally we use our calculator to divide $\ln 1.2375$ by $\ln 1.02$ to get $t \approx 10.76$. So it will take 11 years to get $990 in the account, choice (C).

**Laws of Logarithms:** Here is a review of the basic laws of logarithms.

| Law | Example |
|---|---|
| $\log_b 1 = 0$ | $\log_2 1 = 0$ |
| $\log_b b = 1$ | $\log_6 6 = 1$ |
| $\log_b x + \log_b y = \log_b(xy)$ | $\log_5 7 + \log_5 2 = \log_5 14$ |
| $\log_b x - \log_b y = \log_b(\frac{x}{y})$ | $\log_3 21 - \log_3 7 = \log_3 3 = 1$ |
| $\log_b x^n = n\log_b x$ | $\log_8 3^5 = 5\log_8 3$ |

**More general interest formula:** For interest compounded $n$ times a year we use the formula $A = P\left(1 + \frac{r}{n}\right)^{nt}$ where $n$ is the number of compoundings per year. For example if the interest is being compounded annually (once per year), then $n = 1$, and we get the formula used in the solution above. Other common examples are semiannually ($n = 2$), quarterly ($n = 4$), and monthly ($n = 12$).

# LEVEL 3: ALGEBRA AND FUNCTIONS

69. $(x - y + 3)^2 =$

(A) $(x - y)^2 + 9$
(B) $(x - y)^2 + 9(x - y)$
(C) $(x - y)^2 + 6(x - y) + 9$
(D) $(x - y)^2 + 9(x - y) + 9$
(E) $x^2 - y^2 + 9$

**Solution by picking numbers:** Let's choose values for $x$ and $y$, say $x = 5$ and $y = 2$. Then $(x - y + 3)^2 = (5 - 2 + 3)^2 = 6^2 = \mathbf{36}$. Put a nice big dark circle around **36** so you can find it easier later. We now substitute $x = 5$ and $y = 2$ into each answer choice:

(A) $(5 - 2)^2 + 9 = 18$
(B) $(5 - 2)^2 + 9(5 - 2) = 36$
(C) $(5 - 2)^2 + 6(5 - 2) + 9 = 36$
(D) $(5 - 2)^2 + 9(5 - 2) + 9 = 45$
(E) $5^2 - 2^2 + 9 = 30$

Since A, D, and E each came out incorrect, we can eliminate them. We still have to choose between choices B and C.

Let's choose new values for $x$ and $y$. Setting $x = y = 0$ will work nicely. In this case we have $(x - y + 3)^2 = (0 - 0 + 3)^2 = 3^2 = \mathbf{9}$.

Put a nice big dark circle around **9** so you can find it easier later. We now substitute $x = 0$ and $y = 0$ into each answer choice:

(B) $(0-0)^2 + 9(0-0) = 0$
(C) $(0-0)^2 + 6(0-0) + 9 = 9$

Since B came out incorrect, we can eliminate it and the answer is (C).

**Remark:** Each of the above computations can be done in a single step with your calculator.

* **Algebraic solution:**

$$(x-y+3)^2 = (x-y+3)(x-y+3) = (x-y)^2 + 6(x-y) + 9$$

This is choice (C).

**Notes:** (1) To get from the first expression to the second above, just recall that to square an expression means to multiply that expression by itself.

(2) To get from the second expression to the third we will think of "$x-y$" as a **block**. Formally we can make the substitution $u = x - y$, so that we have

$$(x-y+3)(x-y+3) = (u+3)(u+3) = u^2 + 6u + 9$$
$$= (x-y)^2 + 6(x-y) + 9$$

(3) There are several ways to multiply two binomials. One way familiar to many students is by FOILing. If you are comfortable with the method of FOILing you can use it here to multiply $(u+3)(u+3)$. But an even better way is to use the same algorithm that you already know for multiplication of whole numbers.

Here is how the algorithm works. Try to understand it yourself before reading the full explanation below.

$$\begin{array}{r} u+3 \\ \underline{u+3} \\ 3u+9 \\ \underline{u^2+3u+0} \\ u^2+6u+9 \end{array}$$

What we did here is mimic the procedure for ordinary multiplication. We begin by multiplying 3 by 3 to get 9. We then multiply 3 by $u$ to get $3u$. This is where the first row under the first line comes from.

Next we put 0 in as a placeholder on the next line. We then multiply $u$ by 3 to get $3u$. And then we multiply $u$ by $u$ to get $u^2$. This is where the second row under the first line comes from.

Now we add the two rows to get $u^2 + 6u + 9$.

70. If $k(x) = 3x^5 - 2x^3 + x^2 - x + 1$, for how many real numbers $c$ does $k(c) = 2$ ?

    (A) One
    (B) Two
    (C) Three
    (D) Four
    (E) Five

* **Graphing calculator solution:** On your TI-84 calculator press the Y= button, and enter

$Y_1 = 3X \wedge 5 - 2X \wedge 3 + X \wedge 2 - X + 1$
$Y_2 = 2$

Press ZOOM 6 to see the two graphs in a standard window. Now note that the line $y = 2$ intersects $k(x)$ three times, choice (C).

**Notes:** (1) The points of intersection of $Y_1$ and $Y_2$ are precisely the points where $k(x) = 2$.

(2) If you are having trouble seeing the points of intersection in the standard window, then you can change the window by either using the WINDOW button and adjusting Xmin, Xmax, Ymin, and Ymax (setting the mins to $-5$ and the maxs to 5 works well), or you can press ZOOM 1 to get a ZBOX where you can zoom in on the portion of the graph where the intersection points lie.

71. If $u$ and $v$ are real numbers, $i = \sqrt{-1}$, and

$$(u - v) + 3i = 7 + vi,$$

then what is $u + v$ ?

    (A) 3
    (B) 7
    (C) 10
    (D) 13
    (E) 26

* Two complex numbers are equal if their real parts are equal and their imaginary parts are equal. So we have

$$u - v = 7 \qquad \text{and} \qquad 3 = v.$$

Since $v = 3$ (from the second equation), we have that $u - 3 = 7$. Thus, $u = 7 + 3 = 10$. Finally, $u + v = 10 + 3 = 13$, choice (D).

**Notes:** (1) A **complex number** has the form $a + bi$ where $a$ and $b$ are real numbers and $i = \sqrt{-1}$.

The following are examples of complex numbers:

$2 + 3i$

$\frac{3}{2} + (-2i) = \frac{3}{2} - 2i$

$-\pi + 2.6i$

$0 + 5i = 5i$ This is called a **pure imaginary** number.

$17 + 0i = 17$ This is called a **real number.**

$0 + 0i = 0$ This is **zero**.

$\sqrt{-9} = 3i$

$2 + \sqrt{-2} = 2 + \sqrt{2}\,i$

72. $w(\frac{3}{5t} - \frac{1}{u}) =$

(A) $\frac{2}{5tw-uw}$
(B) $\frac{3uw-5tw}{5ut}$
(C) $\frac{2w}{5tu}$
(D) $\frac{2w}{5t-u}$
(E) $\frac{3w}{5tu}$

**Solution by picking numbers:** Let's let $w = 5$, $t = 2$ and $u = 10$. Then $w\left(\frac{3}{5t} - \frac{1}{u}\right) = 5\left(\frac{3}{10} - \frac{1}{10}\right) = 5\left(\frac{2}{10}\right) = \mathbf{1}$. Put a nice big, dark circle around **1** so you can find it easily later. We now substitute our values for $w$, $t$ and $u$ into each answer choice.

(A) $\frac{2}{50-50}$ = undefined
(B) $\frac{150-50}{100} = 1$
(C) $\frac{10}{100} = 0.1$
(D) $\frac{10}{10-10}$ = undefined
(E) $\frac{15}{100} = 0.15$

Since A, C, D and E are incorrect we can eliminate them. Therefore the answer is choice (B).

* **Algebraic solution:** We have $w\left(\frac{3}{5t} - \frac{1}{u}\right) = w\left(\frac{3u}{5ut} - \frac{5t}{5ut}\right) = \frac{3uw-5tw}{5ut}$. This is choice (B).

**Notes:** (1) To get from the first expression to the second we note that the the **least common denominator** is $5ut$. Since $\frac{3}{5t}$ already has $5t$ in the denominator we only need to multiply the denominator and numerator by $u$ to get $\frac{3}{5t} \cdot \frac{u}{u} = \frac{3u}{5ut}$. Similarly, since $\frac{1}{u}$ already has $u$ in the denominator we only need to multiply the denominator and numerator by $5t$ to get $\frac{1}{u} \cdot \frac{5t}{5t} = \frac{5t}{5ut}$.

(2) To get from the second expression to the third we first rewrite $\frac{3u}{5ut} - \frac{5t}{5ut}$ as $\frac{3u-5t}{5ut}$. We then distribute the $w$ to get $\frac{3uw-5tw}{5ut}$.

73. If $2x + 3y = 5$, $2y + z = 3$, and $x + 5y + z = 3$, then $x =$

(A) 1
(B) 3
(C) 4
(D) 5
(E) 7

* **Solution by performing simple operations:** We add the first two equations and subtract the second equation to get $x = 5 + 3 - 3 = 5$, choice (D).

**Computations in detail:** We add the first two equations:

$$\begin{array}{r} 2x + 3y \qquad = 5 \\ \underline{\qquad 2y + z = 3} \\ 2x + 5y + z = 8 \end{array}$$

We then subtract the third equation from this result:

$$\begin{array}{r} 2x + 5y + z = 8 \\ \underline{x + 5y + z = 3} \\ x \qquad\quad = 5 \end{array}$$

**Solution using Gauss-Jordan reduction:** Push the MATRIX button, scroll over to EDIT and then select [A] (or press 1). We will be inputting a 3 × 4 matrix, so press 3 ENTER 4 ENTER. Then enter the numbers 2, 3, 0 and 5 for the first row, 0, 2, 1 and 3 for the second row, and 1, 5, 1 and 3 for the third row.

Now push the QUIT button (2ND MODE) to get a blank screen. Press MATRIX again. This time scroll over to MATH and select rref( (or press B). Then press MATRIX again and select [A] (or press 1) and press ENTER.

The display will show the following.

$$\begin{array}{l} [[1\,0\,0 \quad 5 \quad\ ] \\ \ \ [0\,1\,0 \ -1.67\,] \\ \ \ [0\,0\,1 \quad 6.33]] \end{array}$$

The first line is interpreted as $x = 5$, choice (D).

**Notes:** (1) In the first paragraph of this solution we created the **augmented matrix** for the system of equations. This is simply an array of numbers which contains the coefficients of the variables together with the right hand sides of the equations.

(2) In the second paragraph we put the matrix into **reduced row echelon form** (rref). In this form we can read off the solution to the original system of equations.

**Warning:** Be careful to use the rref( button (2 r's), and not the ref( button (which has only one r).

74. If $\sqrt{7b^3} = 3.27$, then $b$ =

(A) .43
(B) 1.15
(C) 1.53
(D) 1.76
(E) 12.66

* **Algebraic/calculator solution:** We square each side of the equation to get $7b^3 = 10.6929$. We then divide each side of this last equation by 7 to get $b^3 \approx 1.5276$. Finally, we take the cube root of each side of the last equation to get $b \approx 1.15$, choice (B).

**Notes:** (1) To take a cube root in your calculator you can either use the cube root function found in the MATH menu, or raise the number to the $\frac{1}{3}$ power. In this example you would type 1.5276 ^ (1 / 3) ENTER, or even better just use the calculator's previous answer and type ^ (1 / 3) ENTER.

(2) When possible try to get in the habit of using the calculator's previous answer instead of retyping decimal approximations. For example, this problem can be solved by pressing the following sequence of buttons:

3.27 ^ 2 / 7 ENTER ^ (1 / 3) ENTER

75. If –7 and 5 are both zeros of the polynomial $q(x)$, then a factor of $q(x)$ is

    (A) $x^2 - 35$
    (B) $x^2 + 35$
    (C) $x^2 + 2x + 35$
    (D) $x^2 - 2x + 35$
    (E) $x^2 + 2x - 35$

* **Algebraic solution:** $(x + 7)$ and $(x - 5)$ are both factors of $q(x)$. Therefore so is $(x + 7)(x - 5) = x^2 + 2x - 35$, choice (E).

**Note:** There are several ways to multiply two binomials. One way familiar to many students is by FOILing. If you are comfortable with the method of FOILing you can use it here, but an even better way is to use the same algorithm that you already know for multiplication of whole numbers.

$$\begin{array}{r} x + 7 \\ \underline{x - 5} \\ -5x - 35 \\ \underline{x^2 + 7x + 0} \\ x^2 + 2x - 35 \end{array}$$

What we did here is mimic the procedure for ordinary multiplication. We begin by multiplying –5 by 7 to get –35. We then multiply –5 by $x$ to get $-5x$. This is where the first row under the first line comes from.

Next we put 0 in as a placeholder on the next line. We then multiply $x$ by 7 to get $7x$. And then we multiply $x$ by $x$ to get $x^2$. This is where the second row under the first line comes from.

Now we add the two rows to get $x^2 + 2x - 35$.

**Solution by starting with choice C:** We are looking for the expression that gives 0 when we substitute in –7 and 5 for $x$.

Starting with choice (C) we have $5^2 + 2(5) + 35 = 70$. So we eliminate choice (C).

For choice (D) we have $5^2 - 2(5) + 35 = 50$. So we eliminate choice (D).

For choice (E) we have $5^2 + 2(5) - 35 = 0$ and $(-7)^2 + 2(-7) - 35 = 0$. So the answer is (E).

**Notes:** (1) $c$ is a zero of a function $f(x)$ if $f(c) = 0$. For example, 5 is a zero of $x^2 + 2x - 35$ because $5^2 + 2(5) - 35 = 0$.

(2) A **polynomial** has the form $a_n x^n + a_{n-1}x^{n-1} + \cdots + a_1 x + a_0$ where $a_0$, $a_1$,...,$a_n$ are real numbers. For example, $x^2 + 2x - 35$ is a polynomial.

(3) $p(c) = 0$ if and only if $x - c$ is a factor of the polynomial $p(x)$.

76. If $g(x) = x^2 - 3$ and $g(f(2)) = -2$, then $f(x)$ could be

    (A) $x^3 - x^2 + x - 3$
    (B) $x^3 - x^2 - 3$
    (C) $x^3 + x - 3$
    (D) $x^2 + x - 3$
    (E) $x + 3$

* **Solution by starting with choice C:** We start with choice (C) and guess that $f(x) = x^3 + x - 3$. We then have $f(2) = 2^3 + 2 - 3 = 7$ and $g(f(2)) = g(7) = 7^2 - 3 = 46$. This is incorrect so we can eliminate choice (C).

Let's try choice (B) next and guess that $f(x) = x^3 - x^2 - 3$. It follows that $f(2) = 2^3 - 2^2 - 3 = 1$ and so $g(f(2)) = g(1) = 1^2 - 3 = -2$. This is correct. So the answer is choice (B).

77. If $5x^2 - 2x + 3 = \frac{2}{7}(ax^2 + bx + c)$, then $a + b + c =$

(A) 15
(B) 17
(C) 19
(D) 21
(E) 23

* Letting $x = 1$, the left hand side of the equation is $5(1)^2 - 2(1) + 3 = 6$ and the right hand side is $\frac{2}{7}(a(1)^2 + b(1) + c) = \frac{2}{7}(a + b + c)$. So we have that $\frac{2}{7}(a + b + c) = 6$, and so $a + b + c = 6(\frac{7}{2}) = 21$, choice (D).

78. If $y = \frac{2}{3}(x + 7)$ and $z = y + 212$, then which of the following expresses $z$ in terms of $x$ ?

(A) $z = \frac{2}{3}(x + 219)$

(B) $z = \frac{2}{3}(x + 205)$

(C) $z = \frac{2}{3}(x - 7) + 212$

(D) $z = \frac{2}{3}(x + 7) - 212$

(E) $z = \frac{2}{3}(x + 7) + 212$

* Simply substitute $\frac{2}{3}(x + 7)$ in for $y$ in the second equation to get $z = \frac{2}{3}(x + 7) + 212$, choice (E).

**Remark:** This problem can also be solved by picking a number for $x$. I leave it to the reader to solve the problem this way.

# LEVEL 3: GEOMETRY

79. Lines $k$ and $n$ are perpendicular and intersect at (0,0). If line $n$ passes through the point (–3,1), then line $k$ does NOT pass through which of the following points?

(A) (–2,–6)
(B) (–1, –2)
(C) (1,3)
(D) (3,9)
(E) (7,21)

* Line $n$ has slope $\frac{1}{-3} = -\frac{1}{3}$ and therefore line $k$ has slope 3. So an equation of line $k$ is $y = 3x$. Since $-2 \neq 3(-1)$, the point (–1, –2) is NOT on line $k$. So the answer is choice (B).

**Remarks**: (1) Here we have used the slope formula $m = \frac{y_2 - y_1}{x_2 - x_1}$.

(2) If the line $j$ passes through the origin (the point (0, 0)) and the point $(a, b)$ with $a \neq 0$, then the slope of line $j$ is simply $\frac{b}{a}$.

(3) Perpendicular lines have slopes that are negative reciprocals of each other. The reciprocal of $-\frac{1}{3}$ is $-3$. The negative reciprocal of $-\frac{1}{3}$ is 3.

(4) Note that in answer choices A, C, D, and E, the $y$-coordinate of the point is 3 times the $x$-coordinate of the point.

80. The intersection of a plane with a rectangular solid CANNOT be

    (A) empty
    (B) a point
    (C) a line
    (D) an ellipse
    (E) a triangle

* **Solution by process of elimination:** If the plane is parallel to the rectangular solid the intersection can be empty. So we can eliminate choice (A). The plane can also touch a single vertex or a single edge of the rectangular solid, so we can eliminate choices (B) and (C). A diagonal slice through the solid can result in a triangle. So we can eliminate choice (E) and the answer is choice (D).

**Visual explanation:** The figure below shows a rectangular solid and three planes – one with empty intersection (A), one that intersects the solid in a point (B), and one that intersects the solid in a line (C). Can you draw a picture showing an intersection in a triangle?

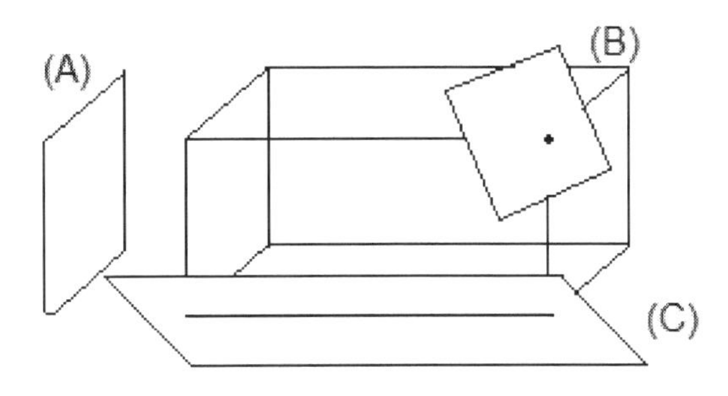

81. What is the distance between the points $(-2,7)$ and $(3,-2)$

(A) 14

(B) $\sqrt{106}$

(C) $\sqrt{26}$

(D) $\frac{9}{5}$

(E) $\frac{5}{9}$

*** Solution using the distance formula:**

$$d = \sqrt{(3-(-2))^2 + (-2-7)^2} = \sqrt{5^2 + (-9)^2} = \sqrt{25+81} = \sqrt{106}$$

This is choice (B).

**Note:** The distance between the points $(s,t)$ and $(u,v)$ is

$$d = \sqrt{(u-s)^2 + (v-t)^2}$$

**Solution using the Pythagorean Theorem:** We plot the two points and form a right triangle

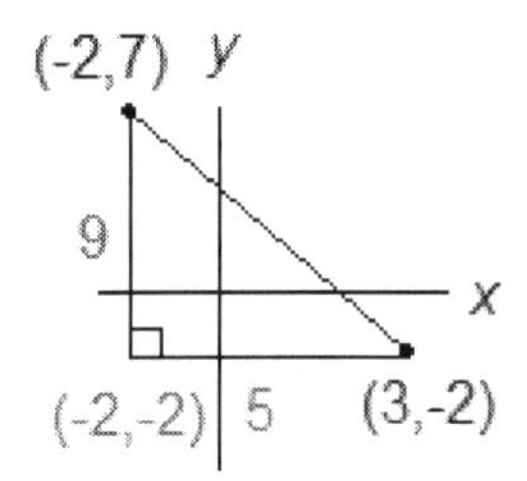


The legs of the triangle have lengths 7 – (–2) = 9 and 3 – (–2) = 5. By the Pythagorean Theorem, the hypotenuse of the triangle has length

$\sqrt{9^2 + 5^2} = \sqrt{106}$, choice (B).

82. What is the surface area of a cube with a volume of 125 $\text{in}^3$

(A) 5 $\text{in}^2$
(B) 25 $\text{in}^2$
(C) 75 $\text{in}^2$
(D) 120 $\text{in}^2$
(E) 150 $\text{in}^2$

* The length of an edge of the cube is $\sqrt[3]{125} = 5$ in. So the surface area of the cube is $6 \cdot 5^2 = 6 \cdot 25 = 150$ in$^2$, choice (E).

**Some formulas:** The **volume of a rectangular solid** is

$$V = lwh,$$

where $l$, $w$ and $h$ are the length, width and height of the rectangular solid, respectively.

In particular, the **volume of a cube** is $V = s^3$ where $s$ is the length of a side of the cube.

The **surface area of a rectangular solid** is just the sum of the areas of all 6 faces. The formula is

$$A = 2lw + 2lh + 2wh$$

where $l$, $w$ and $h$ are the length, width and height of the rectangular solid, respectively.

In particular, the **surface area of a cube** is

$$A = 6s^2$$

where $s$ is the length of a side of the cube.

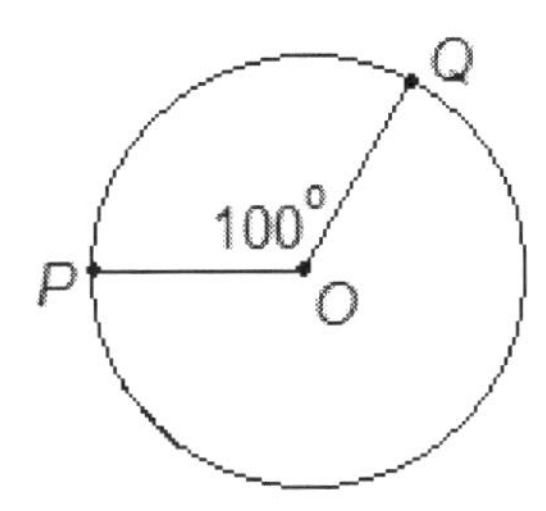


83. If the radius of the circle with center $O$ shown above is 5, what is the length of minor arc $PQ$ ?

    (A) 4.37
    (B) 8.73
    (C) 9.25
    (D) 10.12
    (E) 17.36

* **Solution by setting up a ratio:** The circumference of the circle is $2\pi r = 2\pi(5) = 10\pi$. So we have

$$\frac{100}{360} = \frac{PQ}{10\pi}$$

So $360PQ = 1000\pi$, and therefore $PQ = \frac{1000\pi}{360} \approx 8.73$, choice (B).

**Circle relationship review:**

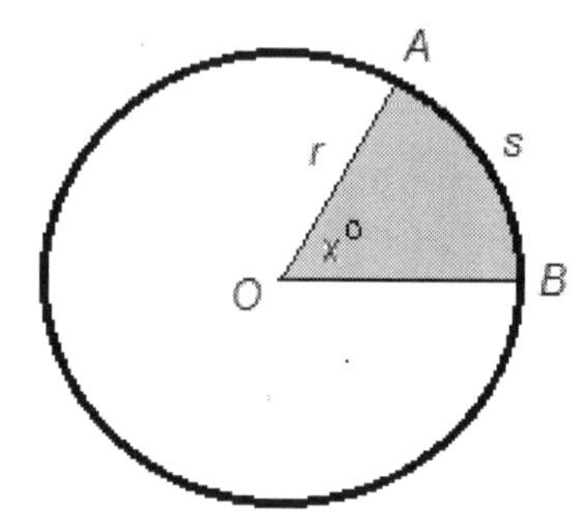


Consider the circle in the figure above. Notice that $\overline{OA}$ and $\overline{OB}$ are both radii of the circle. Therefore $OA = OB = r$. If we know the radius $r$, then we can find the diameter $d$ of the circle, the circumference $C$ of the circle, and the area $A$ of the circle. Indeed, $d = 2r$, $C = 2\pi r$, and $A = \pi r^2$. In fact, if we know any one of the four quantities we can find the other three. For example, if we know that the area of a circle is $A = 9\pi$, then it follows that $r = 3$, $d = 6$, and $C = 6\pi$.

Now, suppose that in addition to the radius $r$, we know the angle $x$. We can then use the following ratio to find the length $s$ of arc $AB$.

$$\frac{x}{360} = \frac{s}{C}$$

We can also use the following ratio to find the area $a$ of sector $AOB$.

$$\frac{x}{360} = \frac{a}{A}$$

84. A cone is inscribed in a right circular cylinder so that the cylinder and cone share a common circular base and the vertex $O$ of the cone is the center of the other circular base. Let $\overline{PQ}$ be a diameter of the circular base common to the cylinder and the cone. If the height of the cylinder is 12 and the diameter of the base is 10, what is the perimeter of triangle $PQO$ ?

    (A) 13
    (B) 18
    (C) 30
    (D) 36
    (E) 60

*** Solution by drawing a picture:**

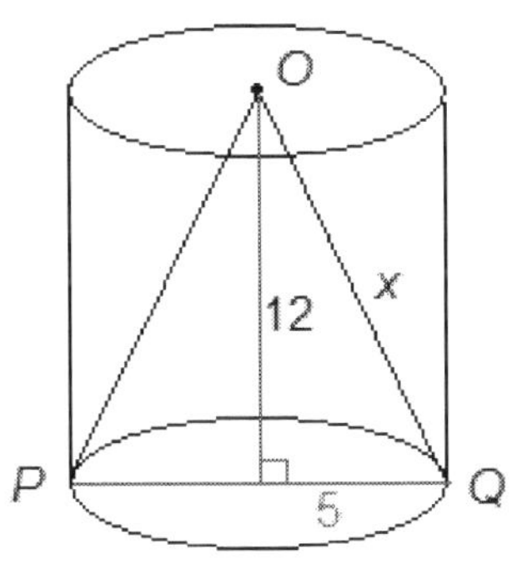


Note that in this picture the radius of a base circle is 5 since the diameter is 10. Using the Pythagorean triple 5–12–13, we have $x = 13$. So the perimeter of triangle $PQO$ is 10 + 13 + 13 = 36, choice (D).

**Notes:** (1) If you do not remember the Pythagorean triple, you can use the Pythagorean Theorem instead.

See problem 30 for more information on Pythagorean triples and the Pythagorean Theorem.

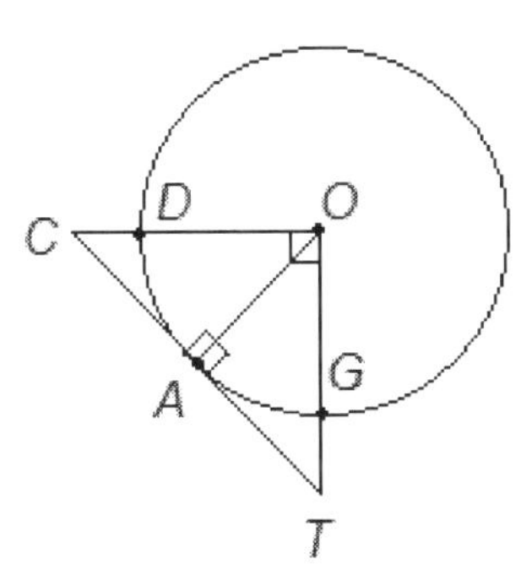


85. The circle in the figure above has center $O$ and a radius of 5. If $OC = OT$, what is $CT + OA + OC - GT$?

    (A) 10
    (B) 15
    (C) 20
    (D) 25
    (E) 30

*** Solution using the fact that the figure is drawn to scale:** We can assume that the figure is drawn to scale (since the figure does not have a note suggesting otherwise). It follows that all three triangles in the figure are isosceles right triangles (or equivalently 45, 45, 90 degree triangles).

So we have $CA = AT = OA = 5$ (since $OA$ is a radius of the circle), $OC = 5\sqrt{2}$, and $GT = OT - OG = 5\sqrt{2} - 5$.

Finally we have

$$CT + OA + OC - GT = (5 + 5) + 5 + 5\sqrt{2} - (5\sqrt{2} - 5) = 20.$$

This is choice (C).

**Notes:** (1) Two triangles are **similar** if their angles are congruent. Note that similar triangles **do not** have to be the same size.

(2) Consider the following figure:

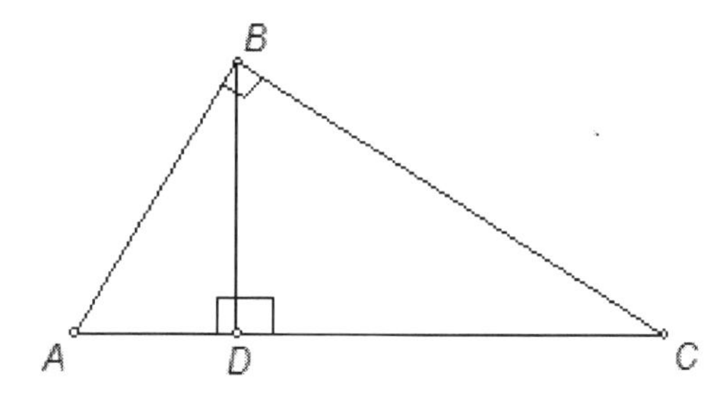


We have a right triangle with an **altitude** drawn from the right angle to the hypotenuse. In this figure triangles $BDC$, $ADB$ and $ABC$ are similar to each other.

In problem 85, we are given that $OC = OT$. This means that triangle $OCT$ is an isosceles right triangle, or equivalently, a 45, 45, 90 triangle. So all three triangles, being similar, are 45, 45, 90 triangles.

(3) You should know the following 45, 45, 90 triangle:

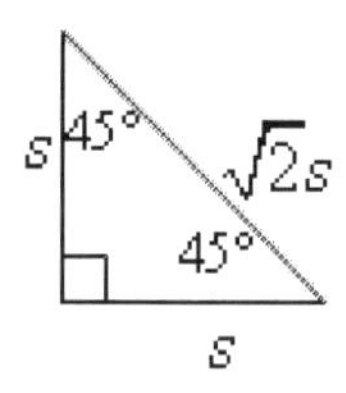


See problem 23 for more details.

(4) The Pythagorean Theorem can be used here as an alternative to using a 45, 45, 90 triangle. I leave the details to the reader.

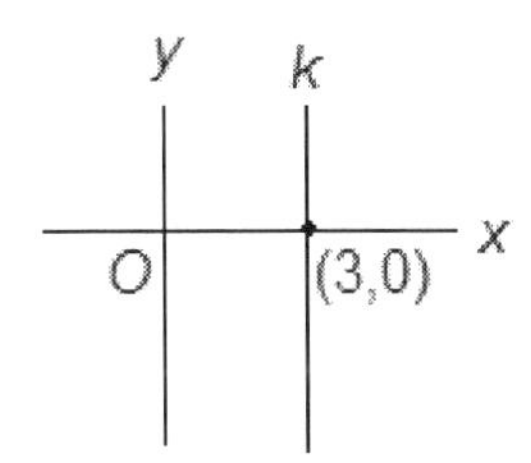


86. An equation of line $k$ in the figure above is

(A) $x = 3$
(B) $y = 3$
(C) $x = 0$
(D) $y = x + 3$
(E) $x + y = 3$

* A vertical line has equation $x = c$, where $c$ is the $x$-coordinate of ANY point on the line. So the answer is $x = 3$, choice (A).

**Note:** Any equation of the form $y = a$ for some real number $a$ is a horizontal line. Any equation of the form $x = c$ for some real number $c$ is a vertical line. Horizontal lines have a slope of 0 and vertical lines have no slope (or to be more precise, **undefined** slope or **infinite** slope).

87. In the rectangular coordinate system the point $P(a, b)$ is moved to the new point $Q(5a, 5b)$. If the distance between point $Q$ and the origin is $k$, what is the distance between point $P$ and the origin?

(A) $\frac{k}{5}$
(B) $k$
(C) $\frac{5}{k}$
(D) $5k$
(E) $25k$

**Solution by picking numbers:** Let's let $a = 1$ and $b = 2$. Then the points are $P(1,2)$ and $Q(5,10)$ and $k = \sqrt{5^2 + 10^2} = \sqrt{125} \approx 11.180$. The distance between $P$ and the origin is $\sqrt{1^2 + 2^2} = \sqrt{5} \approx$ **2.236**. Put a nice big, dark circle around **2.236** so you can find it easily later. We now substitute our value for $k$ into each answer choice.

(A) $\frac{11.180}{5} \approx 2.236$

(B) $k \approx 11.180$

(C) $\frac{5}{k} \approx .447$

(D) $5k \approx 55.9$

(E) $25k \approx 279.5$

Since B, C, D and E are incorrect we can eliminate them. Therefore the answer is choice (A).

* **Direct solution:** The distance between $Q$ and the origin is

$$k = \sqrt{(5a)^2 + (10b)^2} = \sqrt{25a^2 + 100b^2}. = \sqrt{25(a^2 + b^2)}$$
$$= \sqrt{25}\sqrt{a^2 + b^2} = 5\sqrt{a^2 + b^2}$$

The distance between $P$ and the origin is $\sqrt{a^2 + b^2} = \frac{k}{5}$, choice (A).

**Remark:** These distances can also be computed by plotting points, drawing right triangles, and using the Pythagorean Theorem. See the second solution in problem 81 for details.

88. Which of the following is an equation of the line with an $x$-intercept of (4,0) and a $y$-intercept of (0,–3) ?

(A) $y = \frac{3}{4}x + 4$

(B) $y = \frac{3}{4}x - 3$

(C) $y = -\frac{3}{4}x + 4$

(D) $y = -\frac{3}{4}x - 3$

(E) $y = \frac{4}{3}x - 3$

* **Solution by plugging in points:** We plug in the given points to eliminate answer choices. Since the point (0,–3) is on the line, when we substitute a 0 in for $x$ we should get –3 for $y$.

(A) 4
(B) –3
(C) 4
(D) –3
(E) –3

So we can eliminate choices A and C.

Since the point (4,0) is on the line, when we substitute a 4 for $x$ we should get 0 for $y$.

(B) $\frac{3}{4}(4) - 3 = 0$
(D) $-\frac{3}{4}(4) - 3 = -6$
(E) $\frac{4}{3}(4) - 3 \approx 2.33$

So we can eliminate choices (D) and (E), and the answer is choice (B).

**Algebraic solution:** We write an equation of the line in slope-intercept form. The slope of the line is $\frac{-3-0}{0-4} = \frac{3}{4}$. Since (0, –3) is on the line, we have $b = -3$. So an equation of the line is $y = \frac{3}{4}x - 3$, choice (B).

**Remark:** We could have also gotten the slope geometrically by plotting the two points, and noticing that to get from (0, –3) to (4,0) we need to travel up 3 units and right 4 units. So the slope is

$$m = \frac{rise}{run} = \frac{\mathbf{3}}{\mathbf{4}}.$$

**Note:** See the end of problem 18 for more information on slope and slope-intercept form.

## LEVEL 3: PROBABILITY AND STATISTICS

89. The mean test grade of the 17 students in a geometry class was 66. When Johnny took a make-up test the next day, the mean test grade increased to 67. What grade did Johnny receive on the test?

(A) 83
(B) 84
(C) 85
(D) 86
(E) 87

**Solution by changing averages to sums:** We change the averages (or means) to sums using the formula

Sum = Average · Number

We first average 17 numbers. Thus, the **Number** is 17. The **Average** is given to be 66. So the **Sum** of the 17 numbers is $66 \cdot 17 = 1122$.

When Johnny takes his make-up test, we average 18 numbers. Thus, the **Number** is 18. The new **Average** is given to be 67. So the **Sum** of the 18 numbers is $67 \cdot 18 = 1206$.

So Johnny received a grade of $1206 - 1122 = 84$, choice (B).

90. In a small town, 30 families have cats, 50 families have dogs, and 15 families have both cats and dogs. If 25 families have neither cats nor dogs, how many families live in this town?

    (A) 40
    (B) 60
    (C) 75
    (D) 90
    (E) 120

* **Quick computation:** Total = $30 + 50 - 15 + 25 = 90$, choice (D).

**Further explanation:** We used the formula

$$\textbf{Total} = \boldsymbol{C + D - B + N}$$

where $C$ is the number of families that have cats, $D$ is the number of families that have dogs, $B$ is the number of families that have both, and $N$ is the number of families that have neither. We are given that

$$C = 30, D = 50, B = 15, \text{ and } N = 25.$$

It follows that Total = $30 + 50 - 15 + 25 = 90$, choice (D).

**Solution by drawing a Venn Diagram:** Let's draw a Venn diagram.

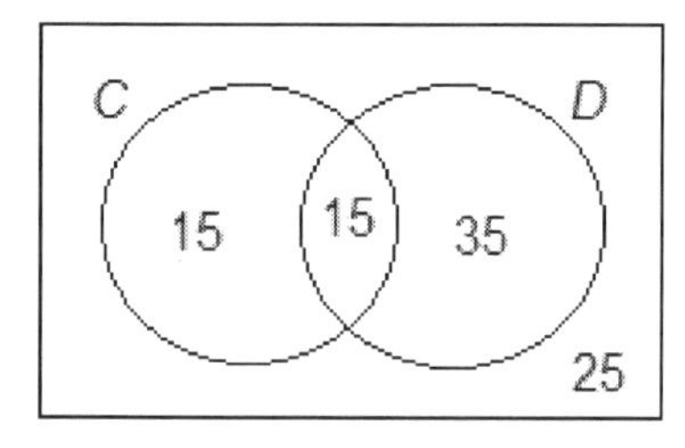


Note that we begin by putting 15 in the middle region. This region is $C \cap D$, the intersection of $C$ and $D$. The 15 is the number of families that have both cats and dogs.

The leftmost 15 is the number of families that have cats only. We get this number by taking the 30 families that own cats and subtracting the number of families that have both cats and dogs. Similarly, the 35 comes from subtracting 15 from 50.

The 25 is the number of families that have neither cats nor dogs.

We get the total number of families by adding all the numbers in this diagram to get 15 + 15 + 35 + 25 = 90, choice (D).

91. When a fair coin is flipped five times, what is the probability that the result will be exactly four heads?

(A) $\frac{1}{32}$
(B) $\frac{3}{32}$
(C) $\frac{5}{32}$
(D) $\frac{7}{32}$
(E) $\frac{9}{32}$

* **Quick solution:** The total number of outcomes is $2^5 = 32$. The number of "successes" is 5. Therefore the probability is $\frac{5}{32}$, choice (C).

**Notes:** (1) To compute a simple probability where all outcomes are equally likely, divide the number of "successes" by the total number of outcomes.

(2) In this problem there are 5 "successes," namely HHHHT, HHHTH, HHTHH, HTHHH, and THHHH.

(3) To compute the total we simply use the counting principle. Each time the coin is flipped there are 2 possibilities (heads or tails). So when the coin is flipped five times there are $(2)(2)(2)(2)(2) = 2^5 = 32$ possibilities. See problem 25 for information on the counting principle.

**Solution using the binomial probability formula:** The probability of an event with probability $p$ occurring exactly $r$ out of $n$ times is

$$\boldsymbol{{}_nC_r \cdot p^r \cdot (1-p)^{n-r}}$$

In this question $n = 5$, $r = 4$, and $p = \frac{1}{2}$. So the desired probability is

$${}_5C_4 \cdot \left(\frac{1}{2}\right)^4 \cdot \left(\frac{1}{2}\right)^1 = 5\left(\frac{1}{2}\right)^5 = \frac{5}{32}, \text{ choice (C).}$$

92. The set $Q$ consists of 15 numbers whose arithmetic mean is zero? Which of the following must also be zero?

I. The median of the numbers in $Q$.
II. The mode of the numbers in $Q$.
III. The sum of the numbers in $Q$.

(A) I only
(B) II only
(C) III only
(D) I and III only
(E) I, II, and III

* **Solution by using a specific list:** Consider the following set:

$$\{-14, 1, 1, 1, 1, 1, 1, 1, 1, 1, 1, 1, 1, 1, 1\}$$

This set has an arithmetic mean of 0, but a median and mode of 1. So I and II do not need to be true. We can therefore eliminate choices A, B, D, and E. So the answer is choice (C).

**Note:** To see that III must be true recall the formula

**Sum = Average · Number**

Since the average (arithmetic mean) is zero, so is the sum.

# LEVEL 3: TRIGONOMETRY

93. In $\Delta DOG$, the measure of $\angle D$ is 60° and the measure of $\angle O$ is 30°. If $\overline{DO}$ is 8 units long, what is the area, in square units, of $\Delta DOG$ ?

(A) 4
(B) 8
(C) $8\sqrt{2}$
(D) $8\sqrt{3}$
(E) 16

* **Solution using a 30, 60, 90 right triangle:** Let's draw two pictures.

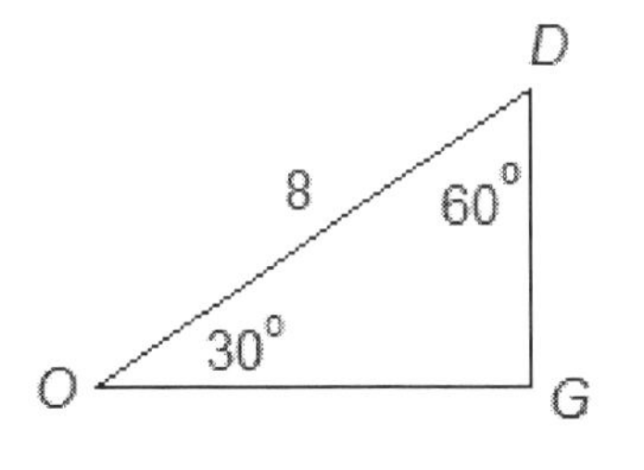


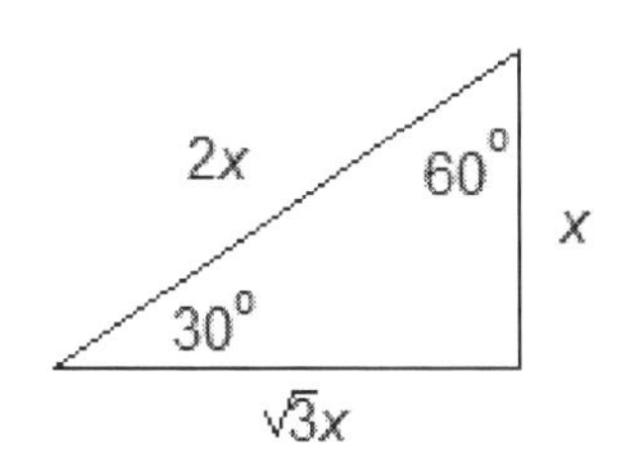


The picture on the left is what is given in the problem. Comparing this to the picture on the right we see that $x = 4$ and $\sqrt{3}x = 4\sqrt{3}$. So the area of the triangle is $\frac{1}{2}(4)(4\sqrt{3}) = 8\sqrt{3}$, choice (D).

**Note:** It is worth memorizing the 30, 60, 90 triangle on the right. You should also commit the 45, 45, 90 triangle to memory (see the end of the solution to problem 23 for details).

**Trigonometric solution:** We have $\sin 30° = \frac{DG}{8}$. So $DG = 8\sin 30°$. Similarly, $\cos 30° = \frac{OG}{8}$. So $OG = 8\cos 30°$. So the area of the triangle is $\frac{1}{2}(OG)(DG) = \frac{1}{2}(8\cos 30°)(8\sin 30°) \approx 13.8564$ (using our calculator). Now plug the answer choices into the calculator and we see that $8\sqrt{3} \approx 13.8564$ choice (D).

**Remark:** Make sure that your calculator is in degree mode. Otherwise you will get the wrong answer.

If you are using a TI-84 (or equivalent) calculator press MODE and on the third line make sure that DEGREE is highlighted. If it is not, scroll down and select it.

94. If $0 < x < 90°$ and $\sin x = 0.525$, what is the value of $\cos(\frac{x}{3})$

    (A) 0.447
    (B) 0.633
    (C) 0.743
    (D) 0.812
    (E) 0.983

* $x = \sin^{-1} 0.525 \approx 31.67$. So $\cos(\frac{x}{3}) \approx \cos\left(\frac{31.67}{3}\right) \approx 0.983$, choice (E).

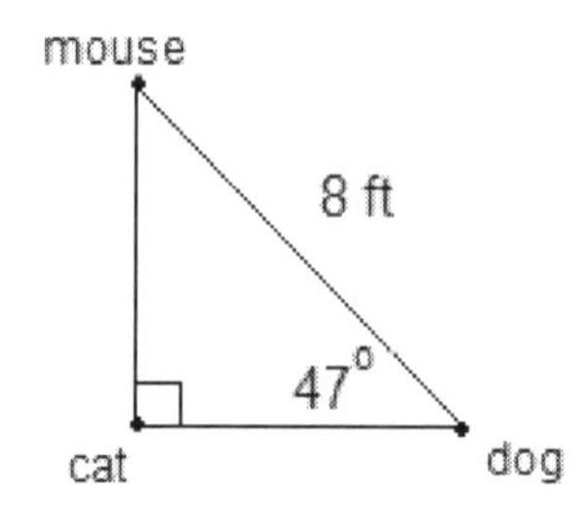


95. A dog, a cat, and a mouse are all sitting in a room. Their relative positions to each other are described in the figure above. Which of the following gives the distance, in feet, from the cat to the mouse?

    (A) 5.5
    (B) 5.9
    (C) 8.57
    (D) 10.9
    (E) 11.7

* Let $x$ be the distance, in feet, from the cat to the mouse. Then we have $\sin 47° = \frac{\text{OPP}}{\text{HYP}} = \frac{x}{8}$. Multiplying each side of the equation $\sin 47° = \frac{x}{8}$ by 8 gives us $8\sin 47° = x$. So $x \approx 5.9$, choice (B).

**Remarks:** (1) If you do not see why we have $\sin 47° = \frac{\text{OPP}}{\text{HYP}}$, review the basic trigonometry given after the solution to problem 30.

(2) If you prefer, you can think of the multiplication above as **cross multiplication** by first rewriting $\sin 47°$ as $\frac{\sin 47°}{1}$.

So we have $\frac{\sin 47°}{1} = \frac{x}{8}$. Cross multiplying yields $8\sin 47° = 1x$. This yields the same result as in the above solution.

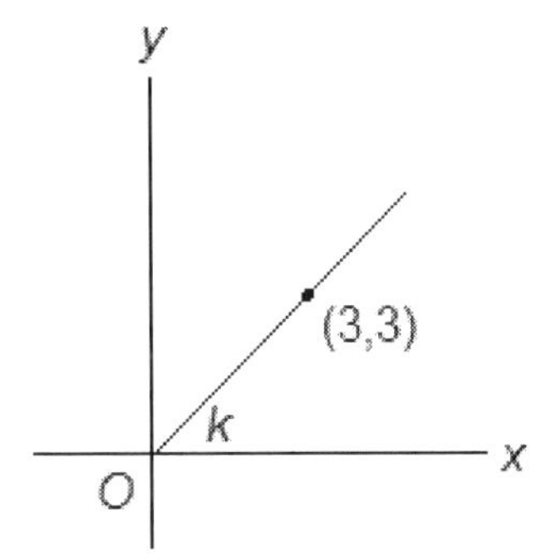


96. In the figure above, $\sin k = ?$

(A) $3\sqrt{2}$

(B) $\frac{\sqrt{2}}{2}$

(C) $\frac{\sqrt{3}}{2}$

(D) 1

(E) $\frac{1}{3}$

* Let's add a little information to the picture.

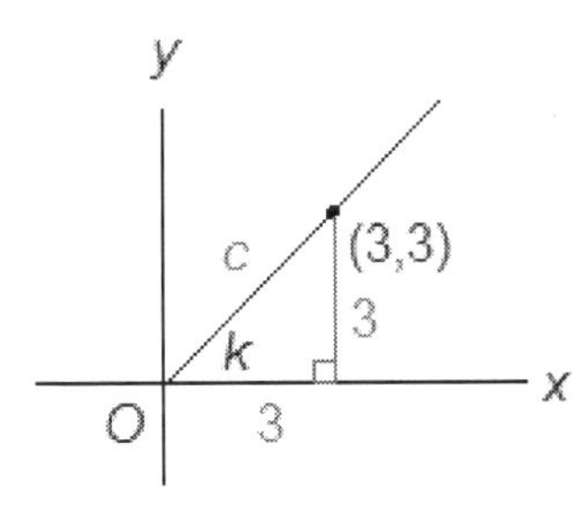


By the Pythagorean Theorem, $c^2 = 3^2 + 3^2 = 9 + 9 = 18$. So $c = \sqrt{18}$.

Now, $\sin k = \frac{OPP}{HYP} = \frac{3}{\sqrt{18}} \sim .707$. We now use our calculator to approximate each answer choice, and we see that $\frac{\sqrt{2}}{2} \sim .707$. So the answer is choice **G.**

**Remarks:** (1) $\sqrt{18}$ can be simplified as $\sqrt{18} = \sqrt{9 \cdot 2} = \sqrt{9}\sqrt{2} = 3\sqrt{2}$. So $\frac{3}{\sqrt{18}} = \frac{3}{3\sqrt{2}} = \frac{1}{\sqrt{2}}$.

Furthermore, we can rationalize the denominator in the expression $\frac{1}{\sqrt{2}}$ to get $\frac{1}{\sqrt{2}} \cdot \frac{\sqrt{2}}{\sqrt{2}} = \frac{\sqrt{2}}{2}$. This is a way to see that the answer is choice G without using our calculator.

(2) Instead of using the Pythagorean Theorem, we can observe that the triangle we formed is an isosceles right triangle which is the same as a 45, 45, 90 right triangle. So the hypotenuse of the triangle has length $3\sqrt{2}$ (see the end of the solution to problem 23 for details).

# LEVEL 4: NUMBER THEORY

97. What number should be added to each of the three numbers –2, 5, and 20 so that the resulting three numbers form a geometric progression?

    (A) 8.125
    (B) 8
    (C) 7.875
    (D) 7.5
    (E) 7

* **Algebraic solution:** We want to find $x$ so that $\frac{5+x}{-2+x} = \frac{20+x}{5+x}$. Cross multiplying gives us $(5+x)^2 = (20+x)(-2+x)$. So we have

$$25 + 10x + x^2 = -40 + 18x + x^2$$
$$25 + 10x = -40 + 18x$$
$$65 = 8x$$
$$\frac{65}{8} = x$$

So $x = \frac{65}{8} = 8.125$, choice (A).

**Notes:** (1) If you are having trouble multiplying the binomials above, see the note at the end of the first solution to problem 69.

(2) A **geometric progression** is a sequence with a **common ratio**. This means that the quotient of any term with the preceding term always gives the same number.

(3) This problem can also be solved by "starting with choice (C)." I leave it to the reader to solve the problem this way.

98. If $a + b \leq a - b$, then $b$ is

(A) positive
(B) negative
(C) nonpositive
(D) nonnegative
(E) zero

* **Algebraic solution:** We subtract $a$ from each side of the inequality to get $b \leq -b$. Now we add $b$ to each side of this last inequality to get $2b \leq 0$. Finally we divide by 2 to get $b \leq 0$, choice (C).

**Solution by picking numbers:** Let's let $a = 2$ and $b = -3$. Then we have $a + b = 2 - 3 = -1$ and $a - b = 2 - (-3) = 5$. Since $-1 \leq 5$ the given condition is satisfied. So we can eliminate choices A, D, and E. If $b = 0$, we get $a \leq a$ which is true. So we can eliminate choice B. Therefore the answer is choice (C).

99. What is the sum of the infinite geometric series?

$\frac{1}{9} - \frac{1}{27} + \frac{1}{81} - \frac{1}{243} + \cdots$ ?

(A) $\frac{1}{81}$
(B) $\frac{1}{27}$
(C) $\frac{1}{15}$
(D) $\frac{1}{12}$
(E) $\frac{1}{6}$

* The sum of an infinite geometric series with first term $a$ and common ratio $r$ with $-1 < r < 1$ is given by $\frac{a}{1-r}$. In this problem we have $a = \frac{1}{9}$ and $r = -\frac{1}{27} \div \frac{1}{9} = -\frac{1}{27} \cdot \frac{9}{1} = -\frac{1}{3}$. So the sum is $\frac{\frac{1}{9}}{1-\left(-\frac{1}{3}\right)} = \frac{\frac{1}{9}}{\frac{4}{3}} = \frac{1}{9} \cdot \frac{3}{4} = \frac{1}{12}$. This is choice (D).

100. If $\log_b 5 = 7$, then $b =$

(A) 1.21
(B) 1.26
(C) 1.47
(D) 1.73
(E) 1.84

* We change the equation to the exponential form $b^7 = 5$ (see Note 2 at the end of problem 66). We now raise each side of this equation to the $\frac{1}{7}$ power to get $b = (b^7)^{\frac{1}{7}} = 5^{\frac{1}{7}} \approx 1.26$, choice (B).

**Notes:** (1) $(b^7)^{\frac{1}{7}} = b^{7 \cdot \frac{1}{7}} = b^1 = b$

(2) For a review of the laws of exponents used here see the end of the solution to problem 4.

# LEVEL 4: ALGEBRA AND FUNCTIONS

101. If $h$ and $k$ are functions, where $h(x) = 1 - 3x + 9x^2 - x^3$ and $k(x) = 51 - 51x + 15x^2 - x^3$, then $k(x) =$

(A) $h(x) + 2$
(B) $h(x) - 2$
(C) $h(x + 2)$
(D) $h(x - 2)$
(E) $2h(x)$

* **Solution by picking a number:** Let's choose a value for $x$, say $x = 2$. Then $k(x) = 51 - 51(2) + 15(2)^2 - 2^3 = \mathbf{1}$. Put a nice big dark circle around **1** so you can find it easier later. We now substitute $x = 2$ into each answer choice:

(A) $h(2) + 2 = 1 - 3(2) + 9(2)^2 - 2^3 + 2 = 25$
(B) $h(2) - 2 = 1 - 3(2) + 9(2)^2 - 2^3 - 2 = 21$
(C) $h(2 + 2) = h(4) = 1 - 3(4) + 9(4)^2 - 4^3 = 69$
(D) $h(2 - 2) = h(0) = 1 - 3(0) + 9(0)^2 - 0^3 = 1$
(E) $2h(x) = 2(1 - 3(2) + 9(2)^2 - 2^3) = 46$

Since A, B, C, and E each came out incorrect, the answer is choice (D).

**Notes:** (1) D is **not** the correct answer simply because it is equal to 1. It is correct because all four of the other choices are **not** 1. **You absolutely must check all five choices!**

(2) To verify this solution algebraically is quite tedious. The computation looks like this:

$$\begin{aligned} h(x - 2) &= 1 - 3(x - 2) + 9(x - 2)^2 - (x - 2)^3 \\ &= 51 - 51x + 15x^2 - x^3 \end{aligned}$$

I leave the details to the reader (only advanced students need to attempt this verification).

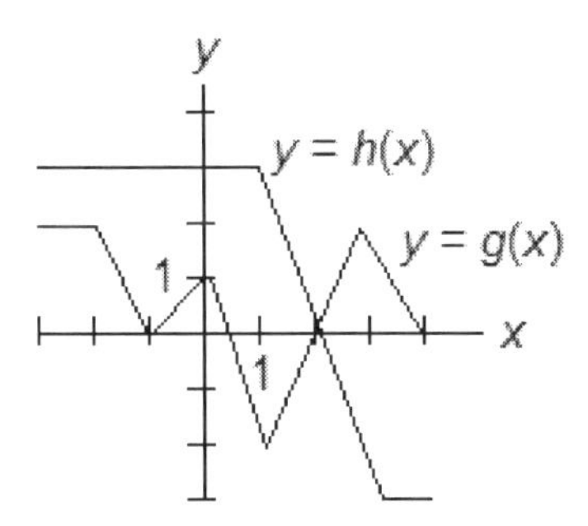


102. The figure above shows the graphs of functions $g$ and $h$. What is the value of $g(h(-3))$ ?

(A) $-2$
(B) $-1$
(C) 0
(D) 1
(E) 2

* $h(-3) = 3$, and so $g\big(h(-3)\big) = g(3) = 2$, choice (E).

**Note:** In general, for a function $f$, we have that $f(a) = b$ is equivalent to "the point $(a, b)$ lies on the graph of $f$."

So for example, $h(-3) = 3$ means that the point $(-3,3)$ lies on the graph of $h$. This can easily be seen from the figure.

Similarly, $g(3) = 2$ means that the point (3,2) lies on the graph of $g$.

103. The product of the two roots of a quadratic equation is $-4$ and their sum is $-2$. Which of the following could be the quadratic equation?

(A) $x^2 - 4x - 2$
(B) $x^2 + 4x - 2$
(C) $x^2 - 2x + 4$
(D) $x^2 + 2x - 4$
(E) $x^2 - 2x - 4$

* In the quadratic equation $\boldsymbol{x^2 + bx + c = 0}$, the product of the roots is $c$, and the sum of the roots is $-b$. So in this problem $c = -4$ and $b = 2$. So the answer is choice (D).

104. If $f(x) = 2x^5 - 1$ and $f^{-1}$ is the inverse function of $f$, what is $f^{-1}(3)$ ?

(A) 0.52
(B) 0.98
(C) 1.15
(D) 1.68
(E) 2.23

**Algebraic solution:** To find $f^{-1}(x)$ we solve for $x$ and then interchange $x$ and $y$.

$$f(x) = 2x^5 - 1$$
$$y = 2x^5 - 1$$
$$y + 1 = 2x^5$$
$$\frac{y+1}{2} = x^5$$
$$\sqrt[5]{\frac{y+1}{2}} = x$$
$$y = \sqrt[5]{\frac{x+1}{2}}$$
$$f^{-1}(x) = \sqrt[5]{\frac{x+1}{2}}$$

Now, $f^{-1}(3) = \sqrt[5]{\frac{3+1}{2}} \approx 1.15$, choice (C).

**Note:** We do that last computation in our calculator. To take the 5th root we can either use the root function in the MATH menu (press 5, then MATH, then 2), or we can raise to the $\frac{1}{5}$ power (type 2 ^ (1/5) ).

* **Solution by starting with choice C:** We are looking for a number $b$ so that $f(b) = 3$. We start with choice (C) and compute

$$f(1.15) = 2(1.15)^5 - 1 \approx 3.$$

So the answer is choice (C).

**Note:** Since the answer did not come out to 3 exactly, you may want to try the other answer choices just to make sure another one doesn't come out closer to 3.

105. If $a, b, c,$ and $d$ are nonzero real numbers and if $a^3b^6c^{-11}d^{14} = \frac{7a^2c^{-11}}{d^{-14}}$ then $ab^6 =$

(A) $\frac{1}{7}$

(B) 7

(C) $7bcd$

(D) $7c^{11}$

(E) $7\frac{c^{11}}{d^{14}}$

* $ab^6 = a^3b^6c^{-11}d^{14} \cdot \frac{c^{11}}{a^2d^{14}} = \frac{7a^2c^{-11}}{d^{-14}} \cdot \frac{c^{11}}{a^2d^{14}} = 7$, choice (B).

**Remarks**: (1) In order to change $a^3b^6c^{-11}d^{14}$ into $ab^6$ we multiply by $c^{11}$ to cancel out $c^{-11}$ and we divide by $a^2$ and $d^{14}$ to cancel out $d^{14}$ in the numerator and to change $a^3$ into $a$.

(2) $c^{-11} \cdot c^{11} = c^{-11+11} = c^0 = 1$. Similarly, $d^{-14} \cdot d^{14} = 1$.

(3) $a^3 \cdot \frac{1}{a^2} = \frac{a^3}{a^2} = a^{3-2} = a^1 = a$.

(4) Take a look at the Law of Exponents table at the end of problem 4.

106. If $5 - \frac{2}{x} = 2 - \frac{5}{x}$, then $\frac{x}{3} =$

(A) $-3$

(B) $-\frac{1}{3}$

(C) 0

(D) $\frac{1}{3}$

(E) $\frac{2}{3}$

* **Algebraic solution:** We add $\frac{5}{x}$ and subtract 5 from each side of the equation to get $\frac{3}{x} = -3$. We then take the reciprocal of each side to get $\frac{x}{3} = -\frac{1}{3}$, choice (B).

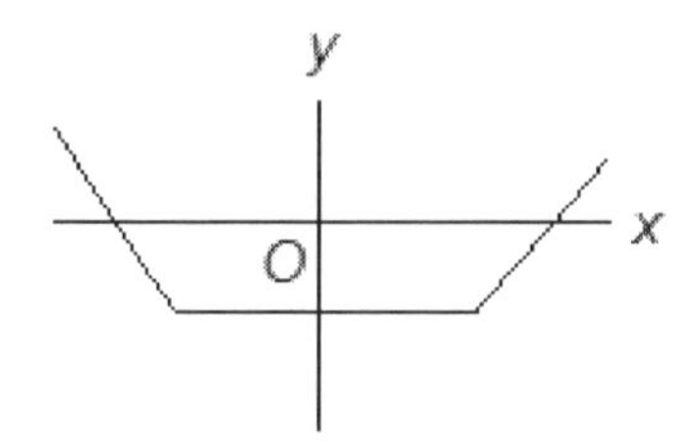


107. The graph of $y = h(x)$ is shown above. Which of the following could be the graph of $y = |h(x)|$ ?

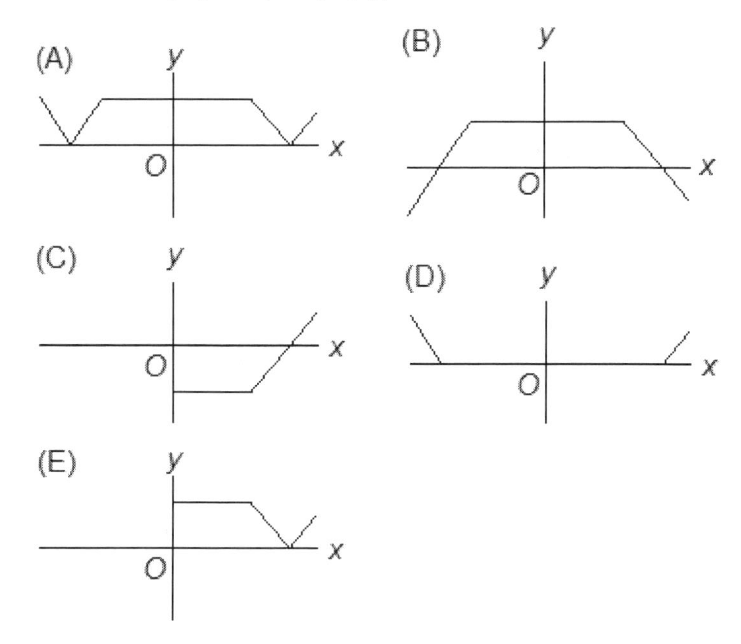


Recall that $|a|$ means the **absolute value** of $a$. It takes whatever number is between the two lines and makes it nonnegative. Here are a few examples: $|3| = 3$, $|-5| = 5$, $|0| = 0$.

* **Graphical solution:** To graph $|h(x)|$ from $h(x)$ simply leave the points that are above (and on) the $x$-axis exactly where they are, and reflect the points that are below the $x$-axis through the $x$-axis. So the answer is choice (A).

108. What is the domain of $k(x) = \sqrt[3]{8 - 12x^2 + 6x - x^3}$ ?

    (A) All real numbers
    (B) $x < -2$
    (C) $-2 < x < 2$
    (D) $x > 2$
    (E) $x > 0$

* The domain of $f(x) = 8 - 12x^2 + 6x - x^3$ is all real numbers since $f(x)$ is a polynomial. The domain of $g(x) = \sqrt[3]{x}$ is also all real numbers. The function $k(x)$ is the composition of these two functions and therefore also has domain all real numbers. So the answer is choice (A).

**Notes:** (1) The **domain** of the function $f(x)$ is the set of allowed $x$-values, or equivalently the set of possible inputs of the function.

(2) Polynomials and cube roots do not cause any problems. In other words you can evaluate a polynomial at any real number and you can take the cube root of any real number.

109. If $r(x) = \frac{7-6x}{3x-4}$, what value does $r(x)$ approach as $x$ gets infinitely larger?

    (A) $-\frac{7}{4}$
    (B) –2
    (C) $\frac{7}{6}$
    (D) $\frac{4}{3}$
    (E) 2

**Calculator solution:** Simply plug in a really large value for $x$ such as 999,999,999. We get (7 – 6 * 999,999,999) / (3 * 999,999,999) – 4) = –2. This is choice (B).

* **Quick solution:** For large $x$, $\frac{7-6x}{3x-4} \approx -\frac{6x}{3x} = -2$, choice (B).

**Detailed solution:** Note that as $x$ gets infinitely large, $\frac{1}{x}$ approaches 0. We now divide each term in both the numerator and denominator by $x$ to get $\frac{7-6x}{3x-4} = \frac{\frac{7}{x}-\frac{6x}{x}}{\frac{3x}{x}-\frac{4}{x}} = \frac{7\left(\frac{1}{x}\right)-6}{3-4\left(\frac{1}{x}\right)}$ which approaches $\frac{7(0)-6}{3-4(0)} = -\frac{6}{3} = -2$, choice (B).

**Remark:** Dividing each term by $x$ is legal because it is equivalent to multiplying the fraction by $\frac{\frac{1}{x}}{\frac{1}{x}} = 1$. We do not have to worry about possibly dividing by 0 because we are assuming that $x$ is very large.

110. If $h(x) = |7 - 2x|$, then $h(-3) =$

(A) $h(-\frac{1}{3})$
(B) $h(\frac{1}{3})$
(C) $h(3)$
(D) $h(7)$
(E) $h(10)$

* We first compute $h(-3) = |7 - 2(-3)| = |7 + 6| = |13| = 13$.

We can now proceed in 2 ways:

Method 1 – Algebraic solution: We want to find the *other* solution of the equation $|7 - 2x| = 13$. So we set $7 - 2x = -13$ and solve for $x$ to get

$$7 - 2x = -13$$
$$-2x = -20$$
$$x = 10$$

So we see that $h(10) = |7 - 2(10)| = |7 - 20| = |-13| = 13$. Therefore the answer is choice (E).

**Remarks:** (1) Recall that $|a|$ means the absolute value of $a$. It takes whatever number is between the two lines and makes it nonnegative.

(2) The equation $|7 - 2x| = 13$ is equivalent to the two equations

$$7 - 2x = 13 \text{ and } 7 - 2x = -13.$$

The solution to the first equation is $-3$ and the solution to the second equation is 10.

Method 2 – Starting with choice C: We start with choice C and compute $h(3) = |7 - 2(3)| = |7 - 6| = |1| = 1$. Since the answer is *not* 13, we can eliminate choice C.

A little thought might suggest that a larger guess is required. If we try choice E next we get $h(10) = |7 - 2(10)| = |7 - 20| = |-13| = 13$. So the answer is choice (E).

# LEVEL 4: GEOMETRY

111. In regular polygon $Q$, the sum of the measures of the interior angles is 3960°. What is the degree measure of one interior angle of $Q$ ?

(A) 60°
(B) 90°
(C) 120°
(D) 165°
(E) 185°

* **Solution using a formula:** The total number of degrees in the interior of an $n$-sided polygon is

$$(n-2)\cdot 180$$

So we have $(n-2)\cdot 180 = 3960$. We divide each side of this equation by 180 to get $n-2=\frac{3960}{180}=22$. So $n=22+2=24$. It follows that $Q$ is a 24 sided polygon. Since $Q$ is regular, the degree measure of one interior angle is $\frac{3960}{24}=165$, choice (D).

**Definition:** A **regular** polygon is a polygon with all sides equal in length, and all angles equal in measure.

112. A sphere is inscribed in a cube with an edge of length 4. What is the volume of the space enclosed by the cube, but NOT by the sphere?

(A) 15.1
(B) 16.3
(C) 18.9
(D) 24.6
(E) 30.5

* The volume of the cube is $V = s^3 = 4^3 = 64$. The radius of the sphere is half the length of an edge of the cube. So $r = 2$ and therefore the volume of the sphere is $V=\frac{4}{3}\pi r^3=\frac{4}{3}\pi(2)^3=\frac{32\pi}{3}$. The desired volume is then $V = 64-\frac{32\pi}{3}\approx 30.5$, choice (E).

**Remark:** Here is a picture of the sphere inscribed in the cube.

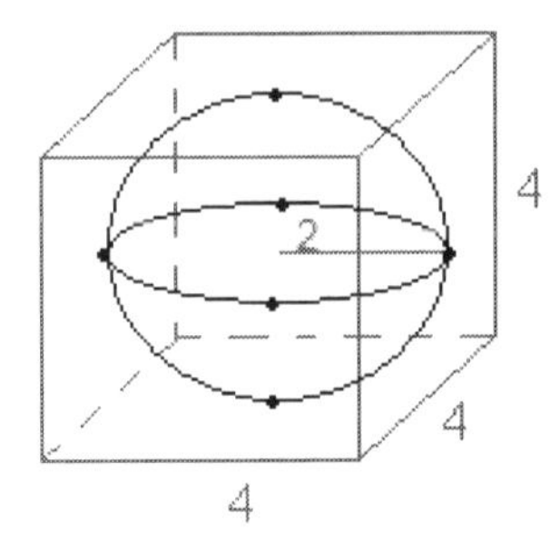


113. Which of the following is an equation of a line perpendicular to $3x + 2y = 5$ ?

(A) $y = -\frac{3}{2}x + \frac{5}{2}$

(B) $y = -\frac{3}{2}x + 1$

(C) $y = \frac{3}{2}x + 1$

(D) $y = \frac{2}{3}x + 1$

(E) $y = -\frac{2}{3}x + 1$

* We first find the slope of the given line by putting it into slope-intercept form. In other words we solve for $y$.

$$3x + 2y = 5$$
$$2y = -3x + 5$$
$$y = -\frac{3}{2}x + \frac{5}{2}$$

From this last equation we see that the given line has a slope of $-\frac{3}{2}$. So the slope of a line perpendicular to this one is $\frac{2}{3}$. Therefore the answer is choice (D).

**Notes:** (1) See the end of problem 18 for more information on slope and slope-intercept form.

(2) Perpendicular lines have slopes that are negative reciprocals of each other. The reciprocal of $-\frac{3}{2}$ is $-\frac{2}{3}$. The negative reciprocal of $-\frac{3}{2}$ is $\frac{2}{3}$.

(3) The slope of a line in the **general form** $ax + by = c$ is $-\frac{a}{b}$. If you choose to memorize this fact, you can find the slope of the line given in this problem quickly without first rewriting the equation in slope-intercept form.

In this question $a = 3$ and $b = 2$. So the slope of the line with equation $3x + 2y = 5$ is $-\frac{3}{2}$.

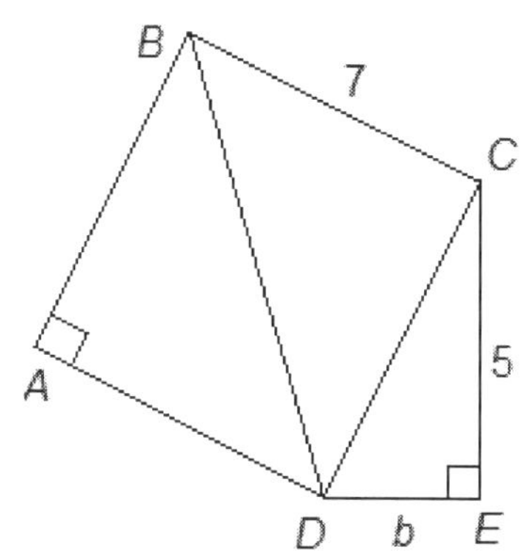


114. In the figure above, $ABCD$ is a square and $\Delta CED$ is a right triangle. What is the value of $b$ ?

  (A) 2
  (B) $2\sqrt{6}$
  (C) 3
  (D) $3\sqrt{3}$
  (E) $3\sqrt{6}$

* **Solution using the Pythagorean Theorem:** Since $ABCD$ is a square, $CD = 7$. By the Pythagorean Theorem, we have $b^2 + 5^2 = 7^2$. So $b^2 + 25 = 49$, and therefore $b^2 = 49 - 25 = 24$. So

$$b = \sqrt{24} = \sqrt{4 \cdot 6} = \sqrt{4}\sqrt{6} = 2\sqrt{6}.$$

This is choice (B).

**Note:** Instead of simplifying $\sqrt{24}$, you can simply approximate it in your calculator, and compare it to similar approximations of the answer choices.

115. In the $xy$-plane, which of the following are the points of intersection of the circles whose equations are $x^2+y^2=7$ and $x^2+(y+1)^2=10$ ?

(A) $(-\sqrt{6},-1)$, $(\sqrt{6},1)$
(B) $(-\sqrt{6},1)$, $(\sqrt{6},-1)$
(C) $(\sqrt{6},1)$, $(\sqrt{6},-1)$
(D) $(-\sqrt{6},1)$, $(-\sqrt{6},-1)$
(E) $(-\sqrt{6},1)$, $(\sqrt{6},1)$

* **Solution by starting with choice C:** We simply check which points in the answer choices satisfy both equations. Starting with choice C, we first check the point $(\sqrt{6},1)$. For the first equation we have

$$\left(\sqrt{6}\right)^2+1^2=7$$
$$6+1=7$$
$$7=7$$

And for the second equation

$$(\sqrt{6})^2+(1+1)^2=10$$
$$6+4=10$$
$$10=10$$

So the point $(\sqrt{6},1)$ is a point of intersection of the two circles.

Let's check the point $(\sqrt{6},-1)$. For the first equation we have

$$\left(\sqrt{6}\right)^2+(-1)^2=7$$
$$6+1=7$$
$$7=7$$

And for the second equation

$$(\sqrt{6})^2+(-1+1)^2=10$$
$$6+0=10$$
$$6=10$$

This last equation is false. So the point $(\sqrt{6},-1)$ is not a point of intersection of the two circles. We can therefore eliminate choice C, as well as choice B and D. (We eliminate D because $\left(\sqrt{6},1\right)$ is not included)

Let's try choice E next and check the point $(-\sqrt{6},1)$. For the first equation we have

$$(-\sqrt{6})^2 + 1^2 = 7$$
$$6 + 1 = 7$$
$$7 = 7$$

And for the second equation

$$(-\sqrt{6})^2 + (1+1)^2 = 10$$
$$6 + 4 = 10$$
$$10 = 10$$

So the point $(-\sqrt{6},1)$ is a point of intersection of the two circles, and the answer is choice (E).

* **Quick solution:** Observe that in both equations there is a term of $x^2$. From this we can see that if the point $(a,b)$ is a point of intersection of the two circles, then so is $(-a,b)$. We can therefore eliminate choices A, B, C, and D. So the answer must be choice (E).

**Note:** This problem can also be solved algebraically in several different ways. For example, we can subtract the first equation from the second and solve for $y$. I leave the details of this solution to the reader.

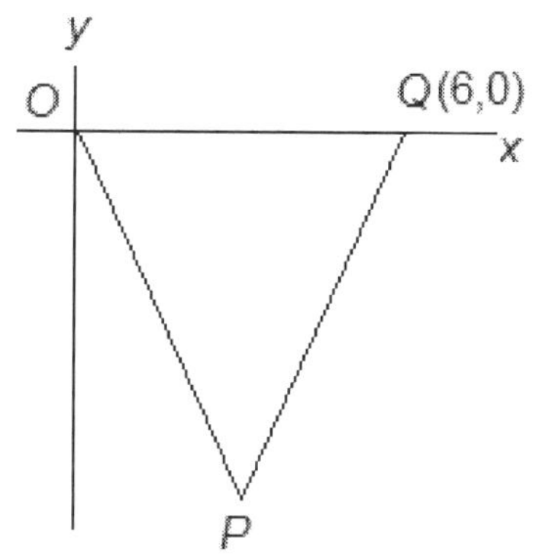


116. In the above triangle, if $OP = QP$ and $m\angle POQ = 60°$, what is the slope of segment $PQ$?

(A) $\frac{1}{\sqrt{3}}$

(B) $\frac{1}{\sqrt{2}}$

(C) $\sqrt{2}$

(D) $\sqrt{3}$

(E) $3\sqrt{3}$

* Let's add some information to the picture.

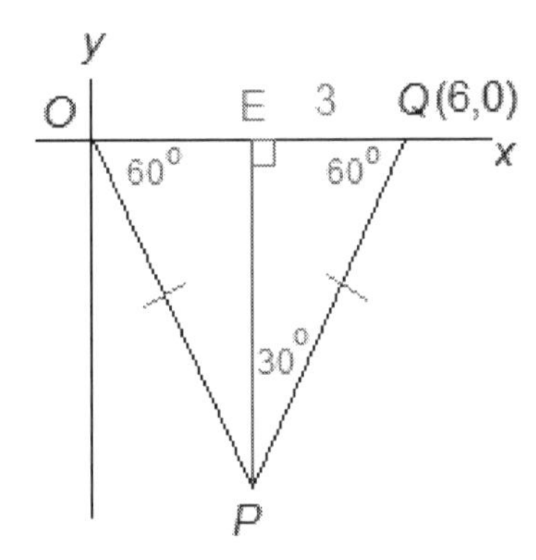


Note the following:

Since $OP = QP$, the triangle is isosceles. So $m\angle PQO = m\angle POQ = 60°$ (and therefore we also have $m\angle OPQ = 60°$) and line segment $\overline{PE}$ is the median, altitude and angle bisector. Since $\overline{PE}$ is a median, $EQ = \frac{1}{2} \cdot 6 = 3$. Since $\overline{PE}$ is an altitude, $m\angle PEQ = 90°$. And since $\overline{PE}$ is an angle bisector, $m\angle EPQ = 30°$.

Since $\Delta PEQ$ is a 30, 60, 90 right triangle, $PE = 3\sqrt{3}$. It follows that the slope of segment $PQ$ is $\frac{\text{rise}}{\text{run}} = \frac{3\sqrt{3}.}{3} = \sqrt{3}$, choice (D).

**Notes:** (1) See the end of problem 18 for more information on slope.

(2) A triangle is **isosceles** if it has two sides of equal length. Equivalently, an isosceles triangle has two angles of equal measure. A triangle is **equilateral** if all three of its sides have equal length. Equivalently, an equilateral triangle has three angles of equal measure (all three angles measure 60 degrees).

(3) An **altitude** of a triangle is perpendicular to the base. A **median** of a triangle splits the base into two equal parts. An **angle bisector** of a triangle splits an angle into two equal parts. In an isosceles triangle, the altitude, median, and angle bisector are all equal (when you choose the base that is **not** one of the equal sides).

(4) For the SAT Math Subject Test, it is worth knowing the following two special triangles:

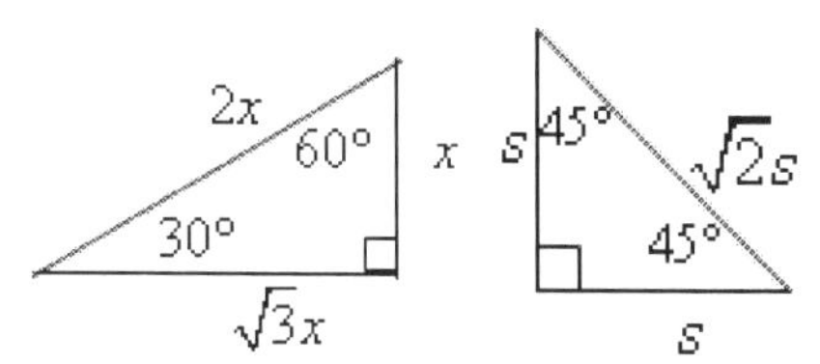


See the end of the solution to problem 23 for details. For this problem, we are using the 30, 60, 90 right triangle and $x = 3$.

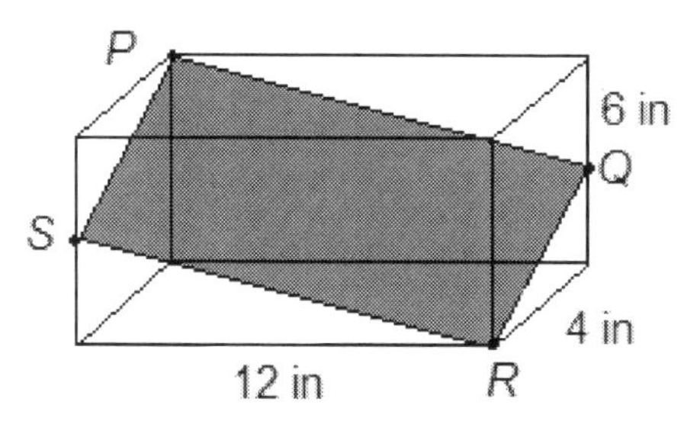


117. The figure above shows a rectangular solid with a length of 12 inches, a width of 4 inches, and a height of 6 inches. If points $S$ and $Q$ are midpoints of edges of the rectangular solid, what is the perimeter of the shaded region?

   (A) 14
   (B) 17.37
   (C) 34.74
   (D) 44.86
   (E) 68.92

* Note that region $PQRS$ is a rectangle. We have $PS = RQ = 5$ since $RQ$ is the hypotenuse of a triangle with legs of lengths 3 and 4 (3, 4, 5 is a Pythagorean triple).

$PQ = SR = \sqrt{3^2 + 12^2} = \sqrt{9 + 144} = \sqrt{153}$ by the Pythagorean Theorem.

So the perimeter of rectangle $PQRS$ is $2(5) + 2\sqrt{153} \approx 34.74$, choice (C).

**Notes:** (1) The most common Pythagorean triples are 3,4,5 and 5, 12, 13. Two others that may come up are 8, 15, 17 and 7, 24, 25.

(2) The Pythagorean Theorem says that if a right triangle has legs of length $a$ and $b$, and a hypotenuse of length $c$, then $c^2 = a^2 + b^2$.

(3) As stated in the solution above $RQ$ is the hypotenuse of a right triangle with legs of lengths 3 and 4. The number 3 comes from the fact that $Q$ is the midpoint of an edge of length 6. Similarly, $S$ is also the midpoint of an edge of length 6.

118. Which of the following is an equation whose graph is the set of points equidistant from the points (2,5) and (7,5) ?

(A) $y = 5$
(B) $x = 5$
(C) $y = 4.5$
(D) $x = 4.5$
(E) $y = x + 4.5$

**Solution by drawing a picture:** Let's draw a picture:

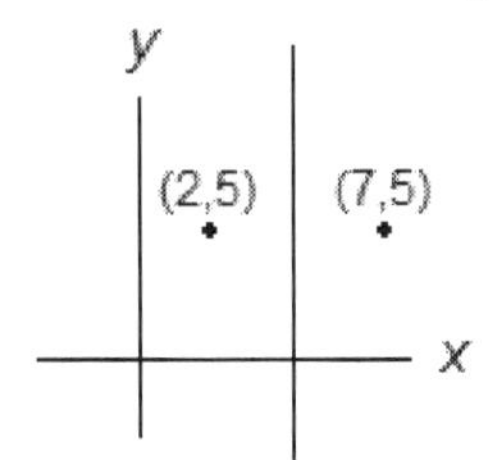


Notice that the set of points equidistant from (2,5) and (7,5) is the perpendicular bisector of the line segment from (2,5) to (7,5) (not shown). The $x$-coordinate of any point on this vertical line is the average of the $x$-coordinates of the two given points: $\frac{2+7}{5} = 4.5$. So the equation of this vertical line is $x = 4.5$, choice (D).

**Note:** Any equation of the form $x = c$ for some real number $c$ is a vertical line. Horizontal lines have a slope of 0 and vertical lines have no slope (or to be more precise, **undefined** slope or **infinite** slope).

* **Quick solution:** The perpendicular bisector of the line segment from (2,5) to (7,5) is the vertical line $x = \frac{2+7}{2} = 4.5$, choice (D).

119. The vertices of rectangle $R$ are (0,0), (0,4), (5,0), and (5,4). Let $S$ be the rectangle that consists of all points $(3x, y - 2)$ where $(x, y)$ is in $R$. What is the area of rectangle $S$ ?

(A) 20
(B) 30
(C) 40
(D) 60
(E) 84

**Solution by drawing a picture:**

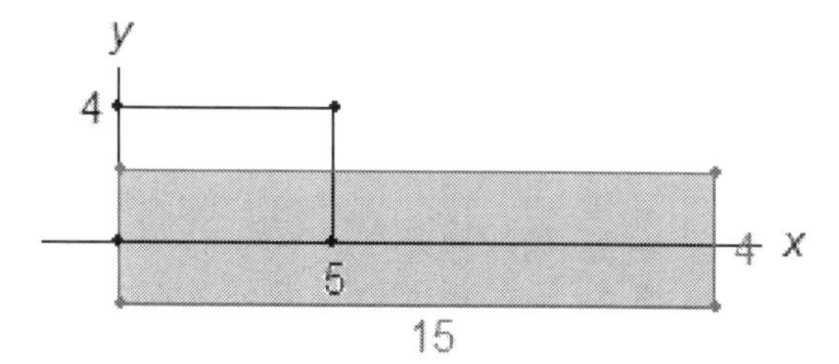


From the picture we see that the length of the new rectangle is 15 and the width is still 4. So the area is (15)(4) = 60, choice (D).

* **Quick solution:** The area of the original rectangle is (5)(4) = 20. We are expanding the rectangle in the $x$ direction by a factor of 3, so the new area is (20)(3) = 60, choice (D).

**Note:** (1) The rectangle is being translated down 2 units. A translation does **not** change the area.

(2) A translation is an example of an **isometry**. An isometry is a transformation that preserves distances. Examples of isometries are translations, rotations and reflections.

(3) Dilations (expansions/contractions) are **not** isometries. In this problem there is a dilation in the $x$ direction. Size is not preserved.

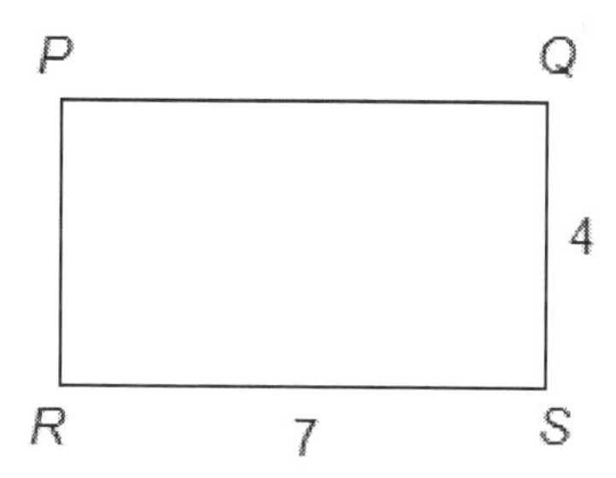


120. What is the volume of the solid generated by rotating rectangle $PQSR$ around $PQ$?

(A) $28\pi$
(B) $49\pi$
(C) $112\pi$
(D) $152\pi$
(E) $196\pi$

* A cylinder is generated with base radius $QS = 4$ and height $RS = 7$. So the volume is $V = \pi r^2 h = \pi(4)^2(7) = 112\pi$, choice (C).

Here is a picture of the cylinder.

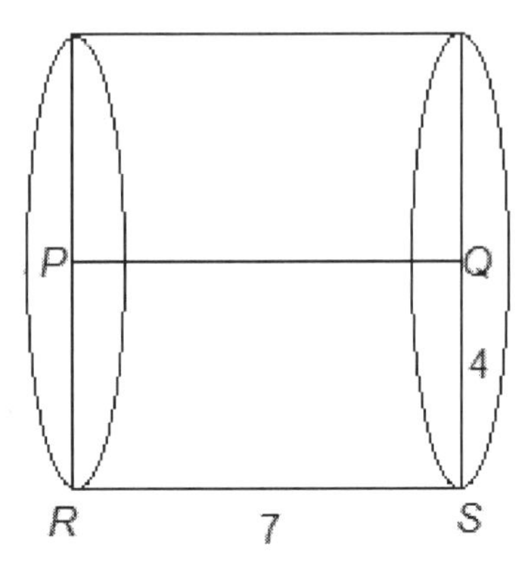


# LEVEL 4: PROBABILITY AND STATISTICS

$$S = 25.33H + 353.16$$

121. The linear regression model above is based on an analysis of the relationship between SAT math scores ($S$) and the number of hours spent studying for SAT math ($H$). Based on this model, which of the following statements must be true?

I. The slope indicates that as $H$ increases by 1, $S$ decreases by 25.33.
II. For a student that studies 15 hours for SAT math, the predicted SAT math score is greater than 700.
III. There is a negative correlation between $H$ and $S$.

(A) I only
(B) II only
(C) III only
(D) I and II only
(E) II and III only

* The slope of the line is $25.33 = \frac{25.33}{1}$. This indicates that as $H$ increases by 1, $S$ increases by 25.33. Also since the slope is positive, there is a **positive correlation** between $H$ and $S$. So I and III are false, and the answer must be choice (B).

**Notes:** (1) We did not have to check II because once we determined that I and III were false, there was only one answer choice left that excluded both of them.

(2) For completeness let's check that II is true. To see this we just need to perform the following computation: $25.33(15) + 353.16 = 733.11 > 700$.

122. A 7-person committee consisting of 3 men and 4 women is to be chosen from a group of 8 men and 7 women. How many different 7-person committees are possible?

 (A) 90
 (B) 91
 (C) 1240
 (D) 1960
 (E) 20108

* There are ${}_8C_3$ ways to choose 3 men from 8, and ${}_7C_4$ ways to choose 4 women from 7. By the counting principle, the total number of 7-person committees is ${}_8C_3 \cdot {}_7C_4 = 1960$.

See problem 26 for more information on the counting principle, and problem 60 for more information on combinations.

123. If $A = \{2, 4, 6, 8, 10\}$, $B = \{-2, 5, 1, -5, k\}$ and $A \cap B = \emptyset$, which of the following CANNOT be the mean of $B$ ?

 (A) 1
 (B) 2
 (C) 3
 (D) 4
 (E) 5

* The sum of the numbers in set $B$ is $-2 + 5 + 1 - 5 + k = k - 1$. Since $A \cap B = \emptyset$, we know that $k$ cannot be 2, 4, 6, 8, or 10. So $k - 1$ cannot be 1, 3, 5, 7, or 9. The sum is $k - 1 = 5$ if and only if the mean of $B$ is $\frac{5}{5} = 1$. So the mean CANNOT be 1, choice (A).

**Alternate solution by plugging in answer choices:** We can also solve this problem by plugging in answer choices. For example, if we guess that the mean of $B$ is 1, then it follows that the sum of the numbers in $B$ is $1 \cdot 5 = 5$. So $-2 + 5 + 1 - 5 + k = 5$, and therefore $k = 6$. But this contradicts $A \cap B = \emptyset$. So the mean cannot be 1, choice (A).

**Notes:** (1) In practice it is normally best to start with choice C here (as opposed to choice A).

(2) To change the mean to a sum we used the formula

**Sum = Average · Number**

124. In Bakerfield, $\frac{1}{3}$ of the population owns at least 1 cat, $\frac{2}{5}$ of the population owns at least 1 dog, and $\frac{1}{2}$ of the population do not own any pets. What fraction of the population own at least 1 cat and 1 dog?

(A) $\frac{29}{30}$

(B) $\frac{1}{2}$

(C) $\frac{3}{10}$

(D) $\frac{7}{30}$

(E) $\frac{1}{5}$

* **Quick computation:** $1 = \frac{1}{3} + \frac{2}{5} - B + \frac{1}{2} = \frac{37}{30} - B$. So $B = \frac{37}{30} - 1 = \frac{7}{30}$, choice (D).

**Notes:** (1) We used the formula

$$\textbf{Total} = \boldsymbol{C + D - B + N}$$

where The Total is 1, $C$ is the fraction of the population that have cats, $D$ is the fraction of the population that have dogs, $B$ is the fraction of the population that have both, and $N$ is the fraction of the population that have neither. We are given that $C = \frac{1}{3}$, $D = \frac{2}{5}$, and $N = \frac{1}{2}$.

(2) This problem can also be solved using a Venn Diagram. See problem 90 to see an example of this type of solution.

* **Solution by picking a specific value for the Total:** If we make the Total 30, then $C = 10$, $D = 12$, and $N = 15$. We can now either use a Venn Diagram or the above formula to get

$$B = C + D + N - \text{Total} = 10 + 12 + 15 - 30 = 7.$$

So the answer is $\frac{7}{30}$, choice (D).

**Note:** 30 is a good choice to pick for the total because it is the least common denominator of the fractions that appear in the problem.

## LEVEL 4: TRIGONOMETRY

125. If $\cos x = 0.36$, then $\cos(180° - x) =$

(A) –0.64
(B) –0.36
(C) 0
(D) 0.36
(E) 0.64

* **Calculator solution:** $x = \cos^{-1} 0.36$, so we can simply type

$$\cos(180 - \cos^{-1} 0.36) \approx -0.36$$

This is choice (B).

**Notes:** (1) The computation above will only give the correct answer if your calculator is in degree mode.

(2) If your calculator is in radian mode you can get the correct answer by using $\pi$ instead of 180.

**Solution using an identity:** We use the following difference identity:

$$\mathbf{\cos(x - y) = \cos x \cos y + \sin x \sin y}$$

$$\cos(180° - x) = \cos 180° \cos x + \sin 180° \sin x$$

$\cos 180° = -1$, $\sin 180° = 0$. So $\cos(180° - x) = -\cos x = -0.36$, choice (B).

126. If $0 \leq x \leq 360°$, $\tan x < 0$ and $\cos x \tan x > 0$, then which of the following is a possible value for $x$ ?

(A) 30°
(B) 90°
(C) 150°
(D) 210°
(E) 330°

* $\tan x < 0$ in Quadrants II and IV. Since $\cos x \tan x > 0$ we must have $\cos x < 0$ . This is true in Quadrants II and III. So $x$ must be in Quadrant II and therefore $90° < x < 180°$. So the answer is choice (C).

**Note:** Many students find it helpful to remember the following diagram.

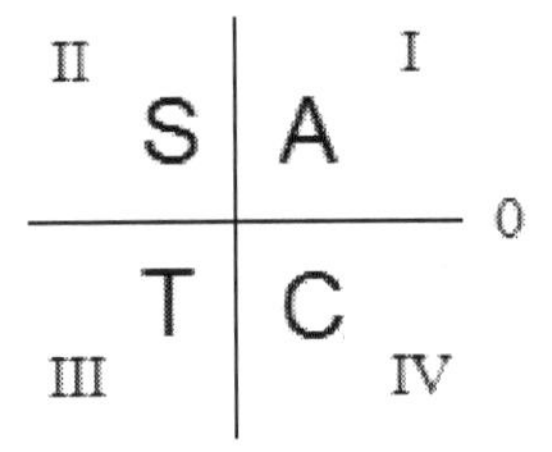


This diagram tells us which trig functions are positive in which quadrants. The **A** stands for "all" so that all trig functions are positive in the first quadrant. Similarly **S** stands for "sine," **T** stands for "tangent," and **C** stands for "cosine."

So, for example, if angle $x$ is in the second quadrant, then $\sin x > 0$, $\tan x < 0$ and $\cos x < 0$.

127. If $\cos x = 0.32$, then what is the value of $\sin x \tan x$ ?

(A) 1.96
(B) 2.12
(C) 2.34
(D) 2.81
(E) 2.98

* $x = \cos^{-1} 0.32$, and so

$$\sin x \tan x = \sin(\cos^{-1} 0.32)\tan(\cos^{-1} 0.32) \approx 2.81.$$

This is choice (D).

128. $\sin x \tan x + \cos x =$

(A) $\sin x$
(B) $\cos x$
(C) $\tan x$
(D) $\cot x$
(E) $\sec x$

*** Solution using basic trig identities:**

$$\sin x \tan x + \cos x = \sin x \frac{\sin x}{\cos x} + \cos x$$

$$= \frac{\sin^2 x}{\cos x} + \frac{\cos^2 x}{\cos x} = \frac{\sin^2 x + \cos^2 x}{\cos x} = \frac{1}{\cos x} = \sec x$$

This is choice (E).

**Notes:** (1) We used the following identities in this solution:

$\tan x = \frac{\sin x}{\cos x}$ $\sec x = \frac{1}{\cos x}$ $\sin^2 x + \cos^2 x = 1$

(2) $\cos x = \cos x \left(\frac{\cos x}{\cos x}\right) = \frac{\cos^2 x}{\cos x}$.

# LEVEL 5: NUMBER THEORY

129. If $\log_b 2 = k$, then $\log_b 32 =$

(A) $5k$
(B) $16k$
(C) $32k$
(D) $k^2$
(E) $k^5$

**Solution by changing to exponential form:** We change the equation to the exponential form $2 = b^k$ (see Note 2 at the end of problem 66). We now raise each side of this equation to the 5th power to get

$$32 = 2^5 = (b^k)^5 = b^{5k}. \text{ So } \log_b 32 = 5k.$$

This is choice (A).

**Note:** See the table at the end of problem 4 for the law of exponents used here.

* **Solution using logarithm laws:** $\log_b 32 = \log_b(2^5) = 5\log_b 2 = 5k$, choice (A).

**Note:** See the table at the end of problem 68 for the law of logarithms used here.

130. A survey was taken from 114 cat owners in Addletown and 225 cat owners in Bergentown. A total of 113 cat owners from the two towns fed their cats "Tasties" cat food. If the same percentage of cat owners from each town used "Tasties," how many of the residents of Bergentown used "Tasties" cat food.

(A) 72
(B) 75
(C) 77
(D) 79
(E) 81

* **Algebraic solution:** Let $x$ be the number of residents from Bergentown that used "Tasties" cat food. It follows that $113 - x$ is the number of residents from Addletown that used "Tasties." Since the same percentage of cat owners from each town used "Tasties" we have

$$\frac{x}{225} = \frac{113-x}{114}$$

We cross multiply to get $114x = 25{,}425 - 225x$.

We now add $225x$ to each side of this last equation to get $339x = 25{,}425$.

We divide each side of this equation by 339 to get $x = 75$, choice (B).

**Notes:** (1) If you do not see where $113 - x$ came from, here is a more detailed explanation: Let $x$ be the number of residents from Bergentown that used "Tasties" cat food, and let $y$ be the number of residents from Addletown that used "Tasties" cat food. We are given that a total of 113 cat owners from the two towns used "Tasties." So $x + y = 113$. It follows that $y = 113 - x$.

(2) Do not forget to check at the end that we have what we were being asked for. After finding $x$, note that we chose $x$ to be the number of residents from Bergentown that used "Tasties." Since this is what is being asked for, $x = 75$ is the answer.

If we had started the problem by letting $x$ be the number of residents from Addletown that used "Tasties," then $x$ would not give the answer. We would need to find $113 - x$.

131. For some real number $a$, the first three terms of an arithmetic sequence are $6a - 3$, $8a + 3$, and $9a$. What is the common difference of the sequence?

(A) 11
(B) 7
(C) –7
(D) –11
(E) –12

* **Algebraic solution:** We want to find $a$ so that

$$(8a + 3) - (6a - 3) = 9a - (8a + 3)$$
$$8a + 3 - 6a + 3 = 9a - 8a - 3$$
$$2a + 6 = a - 3$$
$$a = -9$$

So the first two terms are $6(-9) - 3 = -57$ and $8(-9) + 3 = -69$, and therefore the common difference of the sequence is $-69 - (-57) = -12$, choice (E).

**Remark:** See problem 67 for more information about arithmetic sequences.

132. The only prime factors of the positive integer $k$ are 3, 5, 7, 13, and 23. Which of the following could NOT be a factor of $k$ ?

(A) 15
(B) 21
(C) 91
(D) 143
(E) 299

* **Solution by starting with choice C:** Let's start with choice C. If we start dividing 91 by each of 3, 5, 7, 13, and 23 we see that $91 = 7 \cdot 13$. So we can eliminate choice C.

Trying choice D next we see that $143 = 11 \cdot 13$. Since 143 has a prime factor of 11, 143 cannot be a factor of $k$. So the answer is choice (D).

**Notes:** (1) We are looking for an integer that has a prime factor other than 3, 5, 7, 13 and 23.

(2) Note that $15 = 3 \cdot 5$, $21 = 3 \cdot 7$ and $299 = 13 \cdot 23$. This eliminates choices A, B, and E.

# LEVEL 5: ALGEBRA AND FUNCTIONS

133. What is the greatest integer value of $n$ such that

$$x^2(5n-3) + 2x - 3 = 0$$

has no real roots?

(A) $-2$
(B) $-1$
(C) 0
(D) 1
(E) 2

* The quadratic equation $ax^2 + bx + c$ has no real roots precisely when $b^2 - 4ac$ is negative. In the given equation $a = 5n - 3$, $b = 2$, and $c = -3$. So $b^2 - 4ac$ is equal to

$$2^2 - 4(5n-3)(-3) = 4 + 12(5n-3) = 4 + 60n - 36 = 60n - 32.$$

We need $60n - 32 < 0$, or equivalently $n < \frac{32}{60}$. The greatest such integer value of $n$ is 0, choice (C).

**Remark:** The expression $b^2 - 4ac$ is called the **discriminant** of the quadratic equation. The quadratic equation $ax^2 + bx + c$ has real roots if the discriminant is nonnegative and complex roots if the discriminant is negative.

134. Which of the following lines is an asymptote of the graph of $y = \frac{1-x}{2+x}$?

(A) $x = -1$
(B) $x = 1$
(C) $x = 2$
(D) $y = -1$
(E) $y = \frac{1}{2}$

* **Quick solution:** For large $x$, $y = \frac{1-x}{2+x} \approx \frac{-x}{x} = -1$ So $y = -1$ is a horizontal asymptote of the graph of the given function, choice (D).

**Notes:** (1) The horizontal line with equation $y = b$ is a **horizontal asymptote** for the graph of the function $y = f(x)$ if $y$ approaches $b$ as $x$ gets larger and larger, or smaller and smaller (as in very large in the negative direction).

(2) A **polynomial** has the form $a_n x^n + a_{n-1}x^{n-1} + \cdots + a_1 x + a_0$ where $a_0$, $a_1$,…,$a_n$ are real numbers. If $a_n \neq 0$, then $n$ is the degree of the polynomial.. For example, $1 - x$ is a polynomial of degree 1 (note that this polynomial can be written as $-x^1 + 1$).

(3) A **rational function** is a quotient of polynomials. The function given in this problem is a rational function.

(4) If a rational function consists of polynomials of the **same degree**, then the quick solution above gives a method for finding the only horizontal asymptote of the rational function.

(5) We can also find the horizontal asymptote by plugging in a really large value for $x$ such as 999,999,999. We get

(1 – 999,999,999) / (2 + 999,999,999) = –0.999999997

which is practically $-1$.

(6) The vertical line $x = a$ is a **vertical asymptote** for the graph of the function $y = f(x)$ if $y$ approaches $\pm\infty$ as $x$ approaches $a$ from either the left or right (or both).

(7) If the rational function $y = \frac{p(x)}{q(x)}$ has the property that $q(a) = 0$ and $p(a) \neq 0$, then $x = a$ is a vertical asymptote for the graph.

(8) In this problem plugging in $-2$ makes the denominator of the function 0 and the numerator nonzero. So $x = -2$ is a vertical asymptote for the graph of the given function. Note however that this is not an answer choice.

135. If $h(x) = \frac{x^4 - 16}{x - 2}$ and $k(x) = (x + 2)(x^2 + 4)$, how many points do the graphs of $h$ and $k$ have in common?

(A) None
(B) One
(C) Two
(D) Three
(E) More than three

* **Algebraic solution:**

$$\frac{x^4-16}{x-2}=\frac{(x^2-4)(x^2+4)}{x-2}=\frac{(x-2)(x+2)(x^2+4)}{x-2}=(x+2)(x^2+4),\ x\neq 2$$

So $h(x)=k(x)$ for all $x$-values except for $x=2$, and the answer is (E).

**Notes:** (1) We factored the numerator using the following special factoring formula:

$$(x+y)(x-y)=x^2-y^2$$

(2) This problem can also be solved by picking values for $x$. Once we see that 4 different values give the same result for $h$ and $k$, we would choose choice (E).

136. If $f(x)=ax^3+bx^2+cx+d$ for all real numbers $x$ and if $f(0)=3$ and $f(2)=5$, then $2a+b=$

(A) 0
(B) $\frac{1}{2}$
(C) 2
(D) $\frac{3-c}{4}$
(E) $\frac{1-c}{2}$

* Since $f(0)=3$ we have

$$3=f(0)=a(0)^3+b(0)^2+c(0)+d=d.$$

Therefore $f(x)=ax^3+bx^2+cx+3$.

Since $f(2)=5$ we have

$$5=f(2)=a(2)^3+b(2)^2+c(2)+3=8a+4b+2c+3.$$

Subtracting $2c+3$ from each side of this equation yields

$$2-2c=8a+4b.$$

Finally we divide each side of this equation by 4 to get

$$\frac{1-c}{2}=2a+b.$$

So the answer is choice (E).

137. If $g(x^2 - 2) = x^4 - 4x^2$ for all real numbers $x$, $g(x)$ could be

(A) $x$
(B) $x + 1$
(C) $x + 2$
(D) $x^2 - 2$
(E) $x^2 - 4$

* **Solution by starting with choice C:** If $g(x) = x + 2$, then we have

$$g(x^2 - 2) = (x^2 - 2) + 2 = x^2.$$

So we can eliminate choice (C).

If $g(x) = x^2 - 2$, then

$$g(x^2 - 2) = (x^2 - 2)^2 - 2 = x^4 - 4x^2 + 4 - 2 = x^4 - 4x^2 + 2.$$

So we can eliminate choice (D).

If $g(x) = x^2 - 4$, then

$$g(x^2 - 2) = (x^2 - 2)^2 - 4 = x^4 - 4x^2 + 4 - 4 = x^4 - 4x^2.$$

So the answer is choice (E).

**Note:** (1) $(x^2 - 2)^2 = (x^2 - 2)(x^2 - 2) = x^4 - 2x^2 - 2x^2 + 4$.

(2) See the note at the end of the first solution to problem 69 for an alternative way to multiply polynomials.

138. If $K(x) = \log_5 x$ for $x > 0$, then $K^{-1}(x) =$

(A) $\log_x 5$
(B) $\frac{x}{5}$
(C) $\frac{5}{x}$
(D) $x^5$
(E) $5^x$

* **Quick solution:** The inverse of the logarithmic function $K(x) = \log_5 x$ is the exponential function $K^{-1}(x) = 5^x$, choice (E).

**Notes:** (1) In general, the functions $y = b^x$ and $y = \log_b x$ are inverses of each other. In fact, that is precisely the definition of a logarithm.

(2) The usual procedure to find the inverse of a function $y = f(x)$ is to interchange the roles of $x$ and $y$ and solve for $y$. In this example, the inverse of $y = \log_5 x$ is $x = \log_5 y$. To solve this equation for $y$ we can simply change the equation to its exponential form $y = 5^x$.

(3) For more information on logarithms see problem 66.

139. If $5x - 8y + 2 = 0$ and $x^2 - 4y = 0$ for $x > 0$, then $x =$

(A) 2.23
(B) 2.85
(C) 3.34
(D) 4.76
(E) 5.82

* **Algebraic solution:** If we add $4y$ to each side of the second equation we get $x^2 = 4y$. If we substitute $x^2$ for $4y$ into the left hand side of the first equation we get

$$5x - 8y + 2 = 5x - 2(4y) + 2 = 5x - 2x^2 + 2.$$

So we have $5x - 2x^2 + 2 = 0$ or after multiplying each side of the equation by $-1$ and rearranging terms $2x^2 - 5x - 2 = 0$. We solve for $x$ by using the quadratic formula $x = \frac{-b \pm \sqrt{b^2 - 4ac}}{2a}$.

$$x = \frac{5 \pm \sqrt{(-5)^2 - 4(2)(-2)}}{2(2)} = \frac{5 \pm \sqrt{41}}{4}$$

Since it is given that $x > 0$, we get $x = \frac{5 + \sqrt{41}}{4} \approx 2.85$, choice (B).

**Notes:** (1) There are other ways to solve this probably algebraically. For example, we can multiply the second equation by $-2$ and then add the two equations to get $5x - 2x^2 + 2 = 0$. Now proceed as in the solution above.

(2) This problem can also be solved by plugging in the answer choices. I leave it to the reader to solve the problem this way.

**Graphical solution using a TI-84:** Press Y=, enter (2 + 5X)/8 for $Y_1$ and X^2 / 4 for $Y_2$. Press ZOOM 6 to sketch the graph in a standard window. Now pres CALC (2ND TRACE), select intersect, move the cursor near the positive point of intersection, and press ENTER 3 times. You will see that $x \approx 2.85$, choice (B).

140. If $f$ is a 4th degree polynomial and the points (1,0), (–3,0), $(0,\frac{3}{2})$, and $(2,\frac{175}{6})$ lie on the graph of $f$, then $f(x)$ could equal

(A) $(x-\frac{3}{2})(x-1)(x+\frac{1}{3})(x+3)$

(B) $(x-\frac{3}{2})(x-1)(x+\frac{1}{3})(x+2)$

(C) $(x-\frac{3}{2})(x+1)(x+\frac{1}{3})(x+3)$

(D) $(x-\frac{3}{2})(x+1)(x+\frac{1}{3})(x-3)$

(E) $(x-\frac{1}{3})(x-1)(x+\frac{3}{2})(x+3)$

* Since (1,0) and (−3,0) are points on the graph of $f$, it follows that $x = 1$ and $x = -3$ are zeros of $f(x)$ and so $(x-1)$ and $(x+3)$ are both factors of $f(x)$. Only choices A and E have both of these factors, so we can eliminate choices B, C, and D.

We now plug $x = 0$ into choices A and E to get

(A) $(-\frac{3}{2})(-1)(\frac{1}{3})(3) = \frac{3}{2}$

(E) $(-\frac{1}{3})(-1)(\frac{3}{2})(3) = \frac{3}{2}$

Unfortunately both choices came out correct, so we will have to plug in the last point. We plug $x = 2$ into choices A and E to get

(A) $(2-\frac{3}{2})(2-1)(2+\frac{1}{3})(2+3) = \frac{35}{6}$

(E) $(2-\frac{1}{3})(2-1)(2+\frac{3}{2})(2+3) = \frac{175}{6}$

The answer is choice (E).

**Notes:** (1) The following are all equivalent for a polynomial $f(x)$:

(a) $(c,0)$ is a point on the graph of $f(x)$.
(b) the point $(c,0)$ is an $x$-intercept of the graph of $f(x)$.
(c) $c$ is a zero of $f(x)$.
(d) $(x-c)$ is a factor of $f(x)$.

(2) This problem can be solved just by plugging in points. In other words, choices B,C, and D can be eliminated by plugging in the points (1,0) and (−3,0). This method would be more time consuming however.

141. If $g(-x) = -g(x)$ for all real numbers $x$ and if $(2,-6)$ is a point on the graph of $g$, which of the following points must also be on the graph of $g(x)$ ?

(A) $(-2,-6)$
(B) $(-2,6)$
(C) $(2,6)$
(D) $(-6,2)$
(E) $(6,-2)$

* **Quick solution:** The function $g(x)$ is an odd function. Since the point $(2,-6)$ is on the graph of $g(x)$, so is the point $(-2,6)$, choice (B).

**Remark:** The following are equivalent:

(a) $g(x)$ is an odd function.
(b) $g(-x) = -g(x)$ for all $x$ in the domain of $g$.
(c) The graph of $g(x)$ is symmetrical with respect to the origin.
(d) The point $(-a,-b)$ is on the graph of $g(x)$ whenever the point $(a,b)$ is.
(e) If you rotate the graph of $g(x)$ 180 degrees, the resulting graph is identical to the original.

**Alternate solution:** Since $(2,-6)$ is on the graph of $g(x)$, we have $g(2) = -6$. So $g(-2) = -g(2) = -(-6) = 6$. Therefore $(-2,6)$ is on the graph of $g(x)$, choice (B).

**Note:** $g(a) = b$ is equivalent to "the point $(a,b)$ lies on the graph of $g$."

142. The function $h$, where $h(x) = (2x-3)^2$, is defined for $-3 \leq x \leq 3$. What is the range of $h$?

(A) $0 \leq h(x) \leq 9$
(B) $0 \leq h(x) \leq 81$
(C) $9 \leq h(x) \leq 81$
(D) $-9 \leq h(x) \leq 9$
(E) $-81 \leq h(x) \leq 9$

* Let's begin by trying the extreme $x$-values. $h(3) = (2 \cdot 3 - 3)^2 = 9$, and $h(-3) = (2(-3)-3)^2 = 81$. In particular, 81 is in the range of $h$ and so the answer is either B or C.

Can $h(x) = 0$? Well, if $h(x) = 0$, then $2x - 3 = 0$ and $x = \frac{3}{2}$. So in fact $h\left(\frac{3}{2}\right) = 0$ and therefore the answer is choice (B).

# LEVEL 5: GEOMETRY

143. Let $A$ and $B$ be points in the plane with $A \neq B$. The set all points in the plane that are closer to $B$ than $A$ is

    (A) the interior of a rectangle
    (B) the region in the plane bounded by a hyperbola
    (C) the region in the plane on one side of a line
    (D) the interior of a circle
    (E) the exterior of a circle

* A simple drawing shows that the answer is choice (C).

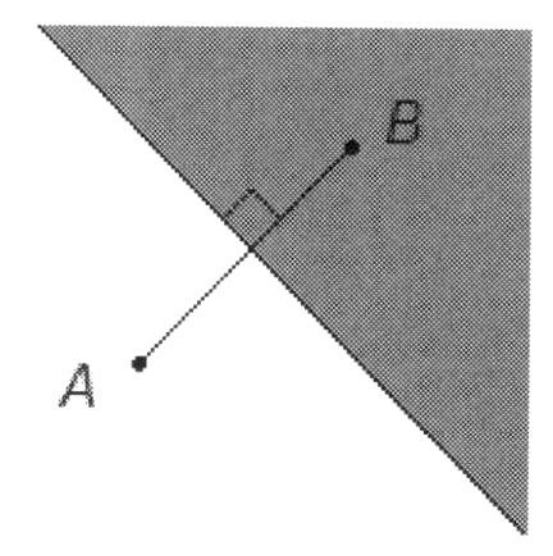


**Notes:** (1) The set of points at the same distance to $A$ and $B$ is the **perpendicular bisector** of the segment $\overline{AB}$.

(2) The set of all points to the right of (and above) the perpendicular bisector is closer to $B$ than $A$. This region is shaded in grey in the figure above.

144. Which of the following could be the coordinates of the center of a circle tangent to the lines $x = 2$ and $y = -1$ ?

    (A) (5,–3)
    (B) (4,–3)
    (C) (3, –3)
    (D) (2,–3)
    (E) (1,–3)

**Solution by drawing a picture:** Let's draw a picture of the two lines.

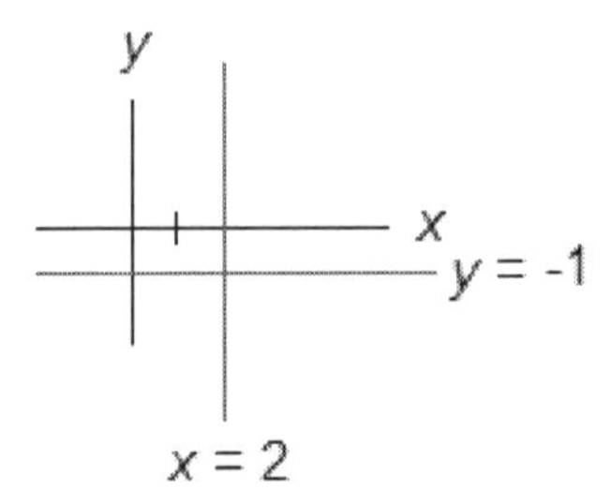


Now note that all of the $y$-coordinates in the answer choices are $-3$. So there are two possible circles that will work.

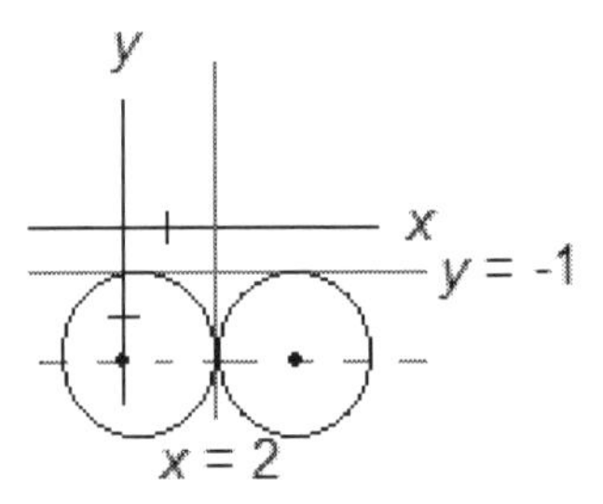


The two circles that would work have centers $(0,-3)$ and $(4,-3)$. So the answer is choice (B).

* **Quick solution:** The $y$-coordinates in the answer choices are all $-3$. So the $y$-coordiante of the center of the circle is two units from the line $y = -1$. Therefore the $x$-coordinate of the center must be two units from the line $x = 2$. So the $x$-coordinate of the center must be 0 or 4, and the answer is choice (B).

145. If $h: (x, y) \rightarrow (x, 3x + y)$ for every pair $(x, y)$ in the plane, for what points $(x, y)$ is it true that $(x, y) \rightarrow (x, y)$ ?

(A) (0,0) only
(B) (0,0) and (0,1) only
(C) The set of points $(x, y)$ such that $y = 0$
(D) The set of points $(x, y)$ such that $y = 1$
(E) The set of points $(x, y)$ such that $x = 0$

* **Algebraic solution:** We want to find all points $(x, y)$ such that $x = x$ and $y = 3x + y$. Subtracting $y$ from each side of the second equation yields $0 = 3x$. Dividing this last equation by 3 gives us $0 = x$. So the answer is choice (E).

**Remark:** This answer can be checked with the following computation $h(0, y) = (0, 3(0) + y) = (0, y)$.

**Solution by plugging in a point:** Let's try the point (0,2). We have $h(0,2) = (0, 3(0) + 2) = (0,2)$. So we can eliminate choices A, B, C, and D. Therefore the answer is choice (E).

**Note:** I used the answer choices as a guide when choosing the point (0,2).

146. Suppose that quadrilateral $PQRS$ has four congruent sides and satisfies $PQ = PR$. What is the value of $\frac{QS}{PR}$?

(A) $\frac{1}{2}$

(B) 1

(C) $\frac{\sqrt{3}}{2}$

(D) $\sqrt{2}$

(E) $\sqrt{3}$

* **Solution by picking a number:** Note that the quadrilateral is a rhombus. Let's draw a picture.

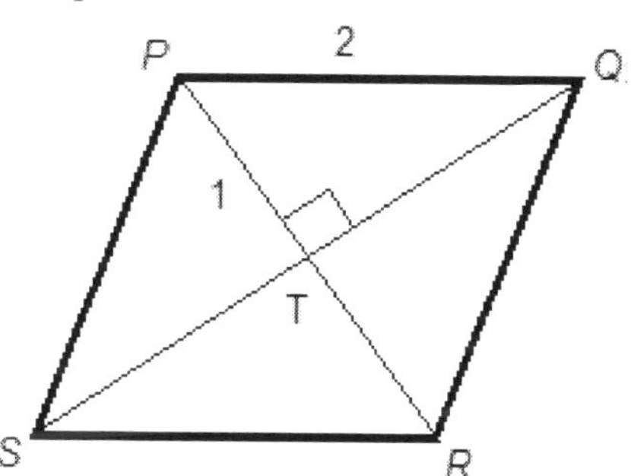


Now, let's choose a value for $PQ$, say $PQ = 2$. Since $PQ = PR$, $PR = 2$ as well. In a rhombus, the diagonals bisect each other, and are perpendicular to each other. It follows that $PT = 1$ and angle $PTQ$ is a right angle. So triangle $PTQ$ is a 30, 60, 90 triangle, and $QT = \sqrt{3}$. Thus, $QS = 2\sqrt{3}$, and it follows that $\frac{QS}{PR} = \frac{2\sqrt{3}}{2} = \sqrt{3}$, choice (E).

**Direct solution:** If we let $PT = x$, then $PQ = 2x$, and by a similar argument to the solution above we have $PR = 2x$ and $QS = 2x\sqrt{3}$. It follows that $\frac{QS}{PR} = \frac{2x\sqrt{3}}{2x} = \sqrt{3}$, choice (E).

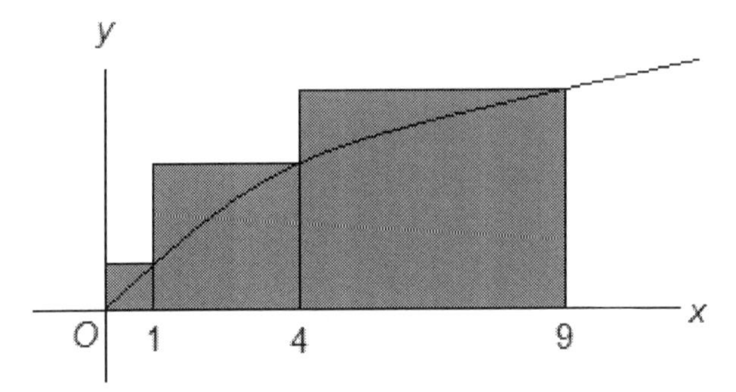


Note: Figure not drawn to scale.

147. The figure above shows a portion of the graph of $y = \sqrt{x}$. What is the sum of the areas of the three circumscribed rectangles shown?

(A) 6
(B) 7
(C) 14
(D) 22
(E) 98

* The leftmost rectangle has a base of 1 and a height of $\sqrt{1} = 1$. So the area of the leftmost rectangle is $1 \cdot 1 = 1$.

The center rectangle has a base of $4 - 1 = 3$ and a height of $\sqrt{4} = 2$. So the area of the center rectangle is $3 \cdot 2 = 6$.

The rightmost rectangle has a base of $9 - 4 = 5$ and a height of $\sqrt{9} = 3$. So the area of the rightmost rectangle is $5 \cdot 3 = 15$.

The sum of these areas is $1 + 6 + 15 = 22$, choice (D).

**Notes:** (1) The length of the interval $[a, b]$ is $b - a$. For example, the length of $[1,4]$ is $4 - 1 = 3$.

(2) The base of each rectangle is the length of an interval. For example, the base of the center rectangle is the length of $[1,4]$ which is $4 - 1 = 3$.

(3) The height of each rectangle is the $y$-coordinate of a point on the graph of the function $y = \sqrt{x}$. More specifically, we use the right endpoint of the interval that forms the base of the rectangle. For example, to get the height of the center rectangle we use the right endpoint of the interval $[1,4]$. So the height is $y = \sqrt{4} = 2$.

148. The height of a right circular cylinder is 3 times the diameter of its base. If the volume of the cylinder is 5, what is the radius of the cylinder?

(A) 0.32
(B) 0.64
(C) 0.76
(D) 0.85
(E) 1.26

* **Algebraic solution:** The diameter of a circle is twice the radius. That is $d = 2r$. Since we are given that the height is 3 times the diameter of the base we have $h = 3d = 3(2r) = 6r$. So we have

$$V = \pi r^2 h$$
$$V = \pi r^2(6r)$$
$$5 = 6\pi r^3$$
$$r^3 = \frac{5}{6\pi}$$
$$r = \sqrt[3]{\frac{5}{6\pi}} \approx 0.64$$

Therefore the answer is choice (B).

**Note:** This problem can also be solved by plugging in the answer choices. I leave it to the reader to solve the problem this way.

149. The intersection of a plane with a cone could be which of the following?

I. A circle
II. A parabola
III. A trapezoid

(A) I only
(B) II only
(C) I and II only
(D) I and III only
(E) I, II, and III

* Let's look at pictures of the four basic conic sections.

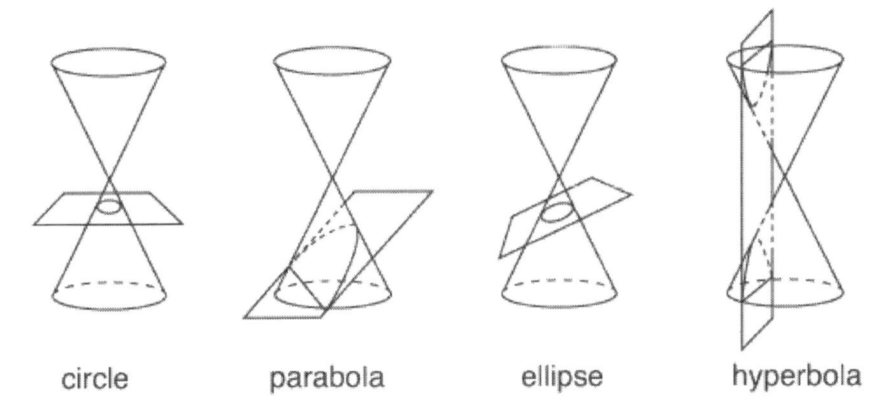


It is now easy to see that the answer is choice (C).

**Note:** Aside from the four standard conic sections shown in the figure above, there are three additional ways a plane can intersect a cone: a point, a line, and a pair of intersecting lines. These are called the **degenerate cases**. I leave it to the reader to draw pictures of these cases.

150. What is the range of the function defined by

$$h(x) = \begin{cases} \sqrt{x+1}, & x > 5 \\ 7 - x, & x \leq 5 \end{cases}?$$

(A) All real numbers
(B) All nonnegative real numbers
(C) All positive real numbers
(D) $2 < y < \sqrt{6}$
(E) $y \geq 2$

*** Solution by graphing:** We sketch a graph of the function.

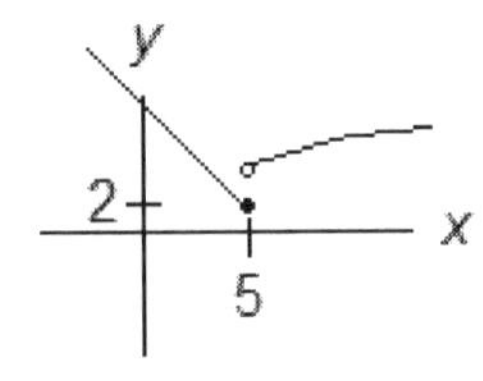


Now note that the minimum $y$ value on the graph is 2, and every $y$ value greater than 2 appears on the graph. So the answer is $y \geq 2$, choice (E).

**Notes:** (1) The **domain** of the function $f(x)$ is the set of allowed $x$-values, or equivalently the set of possible inputs of the function.

Visually we can see the domain on the graph of the function by looking left to right and omitting any "holes." The domain of the function $h(x)$ above is all real numbers.

(2) The **range** of the function $f(x)$ is the set of possible outputs of the function. More specifically it is the set of all $y$-values that have the form $f(x)$ for some $x$ in the domain of the function $f(x)$. Visually we can see the range on the graph of the function by looking up and down and omitting any "holes." Note that in the function $h(x)$ above that below $y = 2$ there are no points on the graph, and for every $y$ value greater than or equal to 2 there is at least one point on the graph (in fact, for $y > \sqrt{6}$ there are 2 points on the graph, and for $2 \leq y < \sqrt{6}$ there is 1 point on the graph).

(3) You can use your graphing calculator to help you draw this graph. Simply graph each function separately (or at the same time) and then copy the functions onto a single set of axes and restrict them appropriately.

(4) Note that the second piece of the function $h(x)$ is a linear equation with slope $m = -1$ and a $y$-intercept of (0,7). See problem 18 for more information about lines.

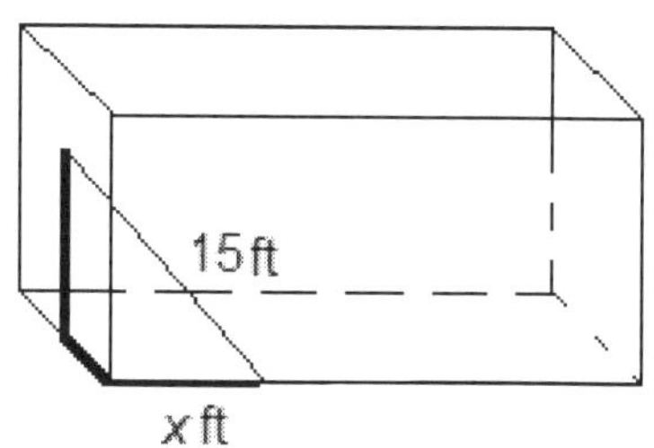


Note: Figure not drawn to scale.

151. A rectangular room measures 20 feet in length and width and 10 feet in height. A fly, with a broken wing, rests at a point 3 feet down from the ceiling at the middle of one end. Some food is located on the floor at a straight line distance of 15 feet from the fly along the edge of the room. The fly walks straight down to the floor, then along the edges of the room straight to the food as shown in the figure above. What is the value of $x$?

(A) 4.23
(B) 8.72
(C) 10.77
(D) 13.27
(E) 73.7

* We use the Generalized Pythagorean Theorem to get

$$15^2 = x^2 + 10^2 + 7^2$$
$$225 = x^2 + 149$$
$$x^2 = 225 - 149 = 76$$
$$x = \sqrt{76} \approx 8.72.$$

This is choice (B).

**Notes:** (1) The **Generalized Pythagorean Theorem** says that the length $d$ of the long diagonal of a rectangular solid is given by

$$d^2 = a^2 + b^2 + c^2$$

where $a$, $b$ and $c$ are the length, width and height of the rectangular solid.

(2) Since the width of the room is 20 feet, the unlabeled horizontal bold line segment has length 10 feet.

(3) Since the height of the room is 10 feet and the fly is 3 feet down from the ceiling, the vertical bold line segment has length 10 – 3 = 7 feet.

152. The circumference of the base of a right circular cone is $10\pi$ and the circumference of a parallel cross section is $8\pi$. If the distance between the base and the cross section is 6, what is the height of the cone?

(A) 12.5
(B) 18
(C) 22
(D) 26
(E) 30

* Let's start by drawing a picture.

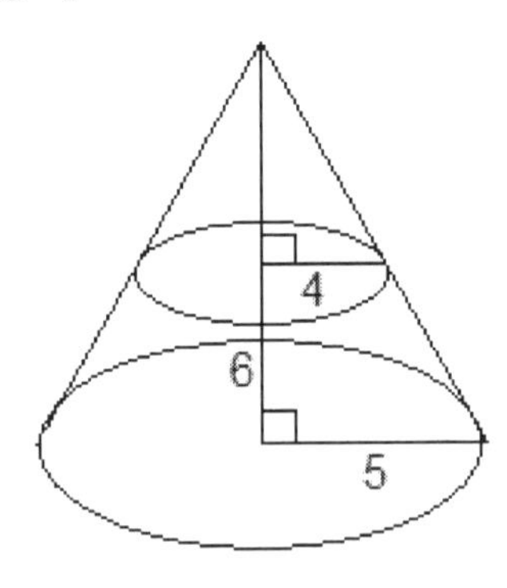

Since the circumference of the base of the cone is $10\pi$, the radius is 5. Similarly, since the circumference of a parallel cross section is $8\pi$, its radius is 4. We now have a pair of similar right triangles. If we let $h$ be the height of the larger triangle, then the height of the smaller triangle is $h - 6$, and we get the ratio

$$\frac{h-6}{h} = \frac{4}{5}$$

Cross multiplying, we get $5h - 30 = 4h$, and so $h = 30$, choice (E).

**Note:** The **circumference** of a circle with radius $r$ is $C = 2\pi r$. So, for example, if the circumference is $C = 10\pi$, then $2\pi r = 10\pi$, and so we have $r = \frac{10\pi}{2\pi} = 5$.

# LEVEL 5: PROBABILITY AND STATISTICS

153. In how many ways can 8 books be split into two piles, one with 5 books and the other with 3 books?

(A) 48
(B) 52
(C) 56
(D) 3136
(E) 6720

* We simply need to count the number of ways to choose 5 books from 8. This is ${}_8C_5 = 56$, choice (C).

**Notes:** (1) See problem 60 for more information on combinations.

(2) Once we choose 5 books from the 8, this automatically also chooses the other pile of 3. A common mistake would be to also compute ${}_8C_3$ and multiply the two numbers together (this is choice D).

(3) We could also compute ${}_8C_3$ instead. In other words we can count the number of ways to choose 3 books from the 8. In this case the pile of 5 will automatically be determined.

(4) Note that in general ${}_nC_r = {}_nC_{n-r}$ . For example, ${}_8C_5 = {}_8C_3$ .

154. Billy has an 8% chance of shooting a basket from the foul line. If Billy makes 3 attempts to make a basket from the foul line, what is the probability that he makes at least 2 shots?

(A) 0.0005
(B) 0.0177
(C) 0.0182
(D) 0.1817
(E) 0.9818

* **Solution using the binomial probability formula:** The probability of an event with probability $p$ occurring exactly $r$ out of $n$ times is

$$\boldsymbol{{}_nC_r \cdot p^r \cdot (1-p)^{n-r}}$$

In this question, $n = 3$, $p = 0.08$ and $r$ can be 2 **or** 3. So the desired probability is

$${}_3C_2 \cdot (0.08)^2 \cdot (0.92)^1 + {}_3C_3 \cdot (0.08)^3 \cdot (0.92)^0 \approx 0.0182$$

This is choice (C).

**Note:** "At least 2 shots" means "2 shots or 3 shots." This is why we have to use the binomial probability formula twice: once for $r = 2$ and once for $r = 3$. We then add the two results.

155. A group of students take a test and the average score is 65. One more student takes the test and receives a score of 92 increasing the average score of the group to 68. How many students were in the initial group?

(A) 5
(B) 6
(C) 7
(D) 8
(E) 9

* **Solution by changing averages to sums:** Let $n$ be the number of students in the initial group. We change the average to a sum using the formula

**Sum = Average · Number**

So the initial **Sum** is $65n$.

When we take into account the new student, we can find the new sum in two different ways.

(1) We can add the new score to the old sum to get $65n + 92$.

(2) We can compute the new sum directly using the simple formula above to get $68(n + 1) = 68n + 68$.

We now set these equal to each other and solve for $n$:

$$65n + 92 = 68n + 68$$
$$24 = 3n$$
$$n = 8.$$

This is choice (D).

156. A positive integer is called a palindrome if it reads the same forward as it does backward. For example, 1661 is a palindrome. If $n$ is a positive integer, how many palindromes are there of length $2n + 1$?

(A) $n$
(B) $10^n$
(C) $10^{n+1}$
(D) $9 \cdot 10^n$
(E) $9 \cdot 10^{2n}$

*** Solution by picking a number:** Let's let $n = 2$, so we are counting the number of palindromes of length $2(2) + 1 = 5$. We use the counting principle. There are 9 possibilities for the leftmost digit, 10 possibilities for the second digit from the left, and 10 possibilities for the middle digit. Since we are counting palindromes, the fourth and fifth digits from the left are determined by the second and first digits, respectively. So there are $9 \cdot 10 \cdot 10 =$ **900** possibilities. We now substitute $n = 2$ into each answer choice.

(A) 2
(B) 100
(C) 1000
(D) 900
(E) 90,000

Since choices A, B, C, and E came out incorrect we can eliminate them and the answer is choice (D).

**Notes:** (1) See problem 26 for more information on the counting principle.

(2) The leftmost digit cannot be 0. That is why there are only 9 possibilities for this digit. All other digits that we choose can be 0, so there are 10 possibilities for each of the rest.

(3) A positive integer of the form $2n + 1$ is odd. Note that we get the integers 3, 5, 7,... when we plug in $n = 1, 2, 3$... and so on.

(4) An odd palindrome of length $2n + 1$ is determined by the first $n + 1$ digits. The first $n$ are to the left of the centermost digit, and the $(n + 1)$st is in the center. The rightmost $n + 1$ must be the same as the leftmost $n + 1$. In particular, note that the rightmost digit cannot be 0 because it must be the same as the leftmost digit which cannot be 0.

# LEVEL 5: TRIGONOMETRY

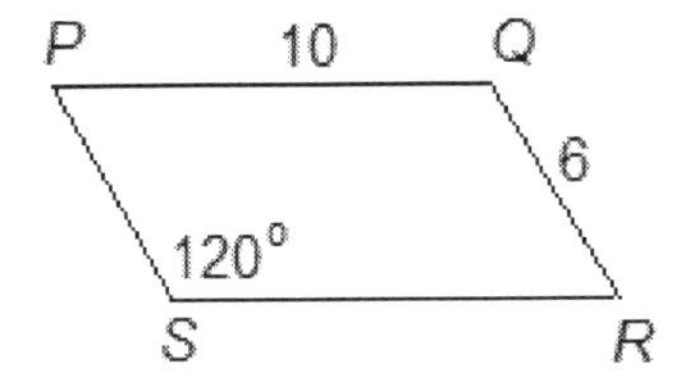


157. The area of parallelogram $PQRS$ shown above is

(A) 30
(B) $30\sqrt{2}$
(C) $30\sqrt{3}$
(D) 60
(E) $60\sqrt{3}$

* The area of the parallelogram is $(6)(10)\sin 120° = 60 \cdot \frac{\sqrt{3}}{2} = 30\sqrt{3}$, choice (C).

**Notes:** (1) We have used the following formula for the area of a parallelogram: $A = ab\sin C$, where $a$ and $b$ are adjacent sides of the parallelogram and $C$ is *any* angle of the parallelogram.

(2) We could have also used 60° for the angle since angle $P$ measures 60°, for example. Note that $\sin 120° = \sin 60°$

(3) If you don't remember how to compute $\sin 120°$ (or $\sin 60°$), you can just use your calculator to get a decimal approximation of $(6)(10)\sin 120°$. Then approximate the answer choices in your calculator to see which answer "matches up."

158. For $0 < x < 90°$,

$$\tan x - \tan(-x) + \sin x - \sin(-x) + \cos x - \cos(-x) =$$

(A) 0
(B) 3
(C) $2\tan x$
(D) $2\tan x + 2\sin x$
(E) $2\tan x + 2\sin x + 2\cos x$

* $\cos x$ is an even function, so that $\cos(-x) = \cos x$. Also, $\sin x$ and $\tan x$ are odd functions, so $\sin(-x) = -\sin x$ and $\tan(-x) = -\tan x$. So we get

$$\begin{gathered}\tan x - \tan(-x) + \sin x - \sin(-x) + \cos x - \cos(-x) \\ = \tan x + \tan x + \sin x + \sin x + \cos x - \cos x \\ = 2\tan x + 2\sin x\end{gathered}$$

This is choice (D).

**Negative Identities:** These identities are just restating what was already described above.

$$\mathbf{\cos(-A) = \cos A} \qquad \mathbf{\sin(-A) = -\sin A}$$

$$\mathbf{\tan(-A) = -\tan A}$$

159. What is the period of the graph of $= \frac{2}{3}\tan(\frac{5}{2}\pi\theta - 2)$ ?

(A) $\frac{4}{15}$
(B) $\frac{2}{5}$
(C) $\frac{2}{3}$
(D) $\frac{4\pi}{15}$
(E) $\frac{2\pi}{5}$

* The period of the graph of $y = a\tan(bx - c)$ is $\frac{\pi}{b}$. So the period of the graph of the given function is $\frac{\pi}{\frac{5\pi}{2}} = \pi \div \frac{5\pi}{2} = \pi \cdot \frac{2}{5\pi} = \frac{2}{5}$, choice (B).

160. If $\cos x = k$, then for all $x$ in the interval $0 < x < 90°$, $\tan x =$

(A) $\frac{1}{1+k}$

(B) $\frac{k}{\sqrt{1+k^2}}$

(C) $\frac{1}{\sqrt{1+k^2}}$

(D) $\frac{\sqrt{1-k^2}}{k}$

(E) $\sqrt{1-k^2}$

* Recall that $\cos x = \frac{\text{ADJ}}{\text{HYP}}$. So we have $k = \frac{k}{1} = \frac{\text{ADJ}}{\text{HYP}}$. Let's draw a picture.

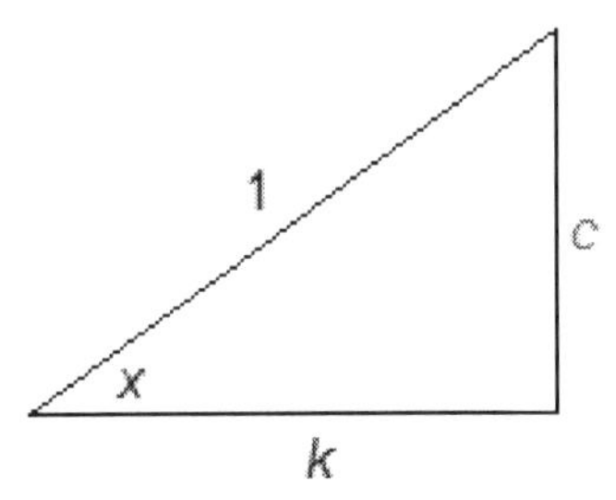


By the Pythagorean Theorem we have $c^2 + k^2 = 1^2$ so that $c^2 = 1 - k^2$ and therefore $c = \sqrt{1-k^2}$. It follows that

$$\tan x = \frac{\text{OPP}}{\text{ADJ}} = \frac{c}{k} = \frac{\sqrt{1-k^2}}{k},$$ choice (D).

**Remark:** If you do not see why we have $\cos x = \frac{\text{ADJ}}{\text{HYP}}$ or $\tan x = \frac{\text{OPP}}{\text{ADJ}}$ review the basic trigonometry given after the solution to problem 30.

# Supplemental Problems Questions

## Level 1: Number Theory

1. If $n$ is an odd integer, then which of the following must be an even integer?

   (A) $n+2$
   (B) $6n+3$
   (C) $n^2+n^5$
   (D) $2n^2+5n^5$
   (E) $4n^2-1$

2. A room has 1200 square feet of surface that needs to be painted. If 2 gallons of paint will cover 450 square feet, what is the least whole number of gallons that must be purchased in order to have enough paint to cover the entire surface?

   (A) 2
   (B) 3
   (C) 4
   (D) 5
   (E) 6

3. What is the smallest positive integer divisible by 100, 70, and 30?

   (A) 210,000
   (B) 2100
   (C) 210
   (D) 180
   (E) 60

4. The number 150 is increased by 35%. This new number is decreased by 35%. What is the final result to the nearest integer?

   (A) 125
   (B) 132
   (C) 150
   (D) 162
   (E) 203

# LEVEL 1: ALGEBRA AND FUNCTIONS

5. If $b = 3$, then $(b - 7)(b + 1) =$

   (A) – 16
   (B) – 4
   (C) – 1
   (D) 7
   (E) 16

6. If $5^x = 7$, then $5^{2x} =$

   (A) 11
   (B) 13
   (C) 25
   (D) 49
   (E) 625

7. If $f(x) = \sqrt{x^2 - 1}$, then $f(-1.8) =$

   (A) 0.5
   (B) 1
   (C) 1.5
   (D) 2
   (E) 2.5

8. Suppose the function $S$, where $S(h) = 25.33h + 353.16$ is used to model the relationship between SAT math subject test scores $S(h)$ and the number of hours $h$ spent studying for the SAT math subject test. Based on this model, a student that received a score of 650 on the SAT math subject test studied how many hours?

   (A) 8.6
   (B) 10.5
   (C) 11.2
   (D) 11.7
   (E) 12.3

9. If $5c^3 = 4$, then $10(5c^3)^2 =$

(A) 54
(B) 80
(C) 96
(D) 120
(E) 160

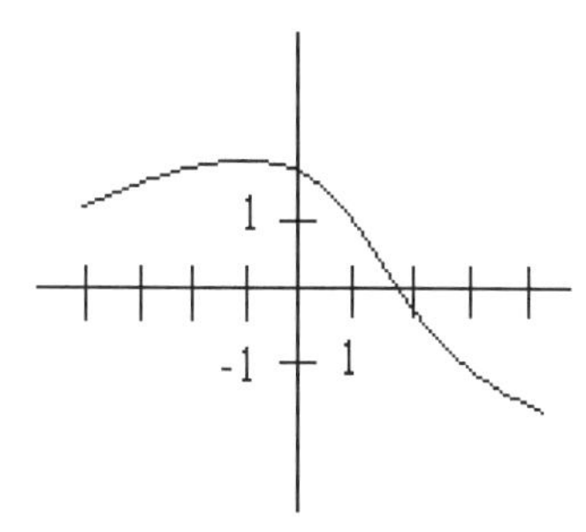


10. The figure above shows the graph of the function $f$ . Which of the following is greater than $f(-2)$?

(A) $f(-4)$
(B) $f(-3)$
(C) $f(-1)$
(D) $f(1)$
(E) $f(2)$

11. If $x^2 - y^2 = 1$ and $x^2 + y^2 = 7$, then $x$ could be

(A) $-1$
(B) 0
(C) 1
(D) 2
(E) 3

12. The equation $x^4 - 3x^3 + 2x^2 + 1 = x^3 - x^2 + 4x + 1$ is equivalent to

(A) $-x^4 + 4x^3 - 3x^2 + 4x = 0$
(B) $x^4 + 4x^3 + 3x^2 + 4x = 0$
(C) $-x^4 - 4x^3 + 3x^2 + 4x = 0$
(D) $x^4 - 4x^3 + 3x^2 + 4x = 0$
(E) $-x^4 + 4x^3 + 3x^2 + 4x + 2 = 0$

13. A group of $x$ children has collected 40 playing cards. If each child collects $y$ more playing cards per week for the next $w$ weeks, which of the following represents the number of playing cards that will be in the group's collection?

(A) $40xy$
(B) $40+\frac{wy}{x}$
(C) $40+\frac{wx}{y}$
(D) $40+xy+w$
(E) $40+wxy$

14. Which of the following expressions is equivalent to $\frac{5k+50}{5}$ ?

(A) $k+10$
(B) $k+50$
(C) $7k+10$
(D) $11k$
(E) $50k$

## LEVEL 1: GEOMETRY

15. The width of a rectangle is three times its length. If the area of the rectangle is 75 inches, what is its perimeter?

(A) 20
(B) 25
(C) 30
(D) 35
(E) 40

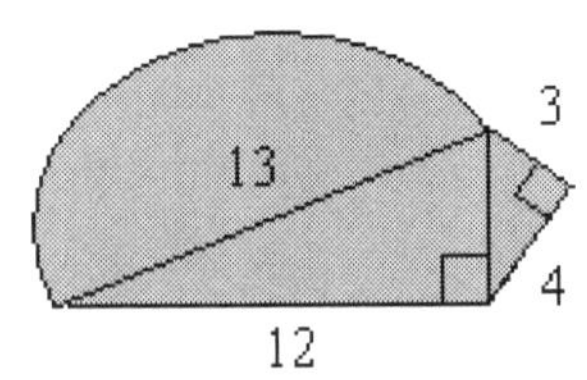


16. What is the total area of the shaded region above to the nearest integer?

(A) 204
(B) 102
(C) 86
(D) 51
(E) 42

17. Which of the following is an equation of the line in the $xy$-plane that passes through the point $(0,-7)$ and is perpendicular to the line $y = -6x + 2$?

(A) $y = -6x + 7$
(B) $y = -6x + 14$
(C) $y = -\frac{1}{6}x + 6$
(D) $y = \frac{1}{6}x - 7$
(E) $y = \frac{1}{6}x + 7$

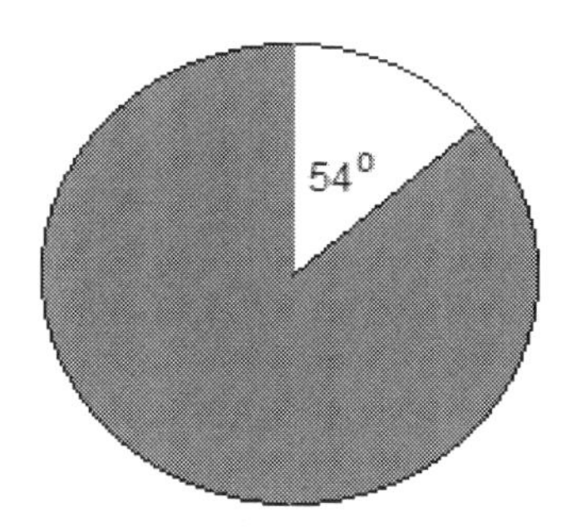


18. In the figure above, what percentage of the circle is shaded?

(A) 65%
(B) 70%
(C) 75%
(D) 80%
(E) 85%

19. If $(4, y)$ is an intersection point of the graphs of $\sqrt{x} = 2$ and $y^2 = \frac{x^3}{256}$, then $y$ could be

(A) $-2$
(B) $-\frac{1}{2}$
(C) $1$
(D) $\frac{3}{2}$
(E) $2$

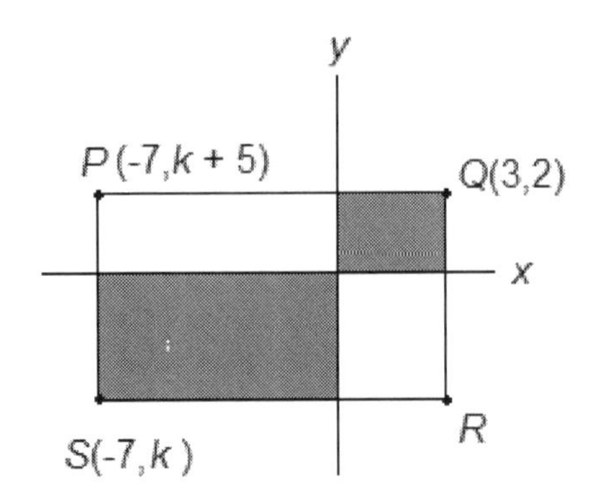


20. In the figure above, $PQRS$ is a rectangle. What is the area of the shaded region?

(A) 6
(B) 21
(C) 27
(D) 35
(E) 50

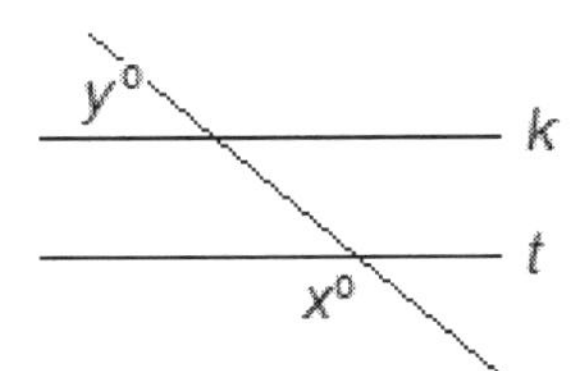


21. In the figure above, lines $k$ and $t$ are parallel. What is the value of $x + y$?

(A) 360°
(B) 270°
(C) 180°
(D) 90°
(E) It cannot be determined from the given information.

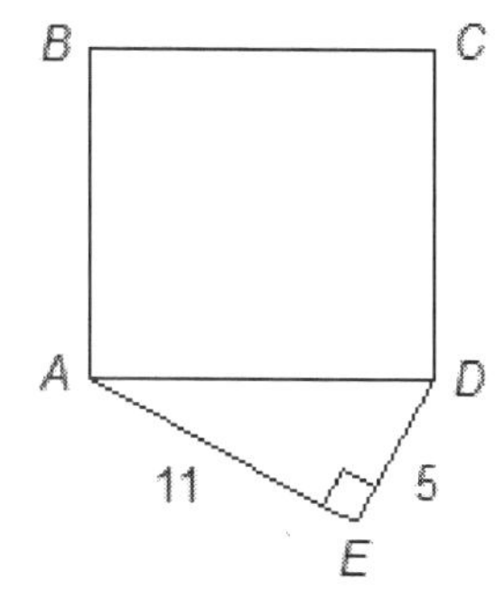


22. In the figure above, what is the area of square $ABCD$ ?

(A) 12
(B) 27.5
(C) 55
(D) 146
(E) 160

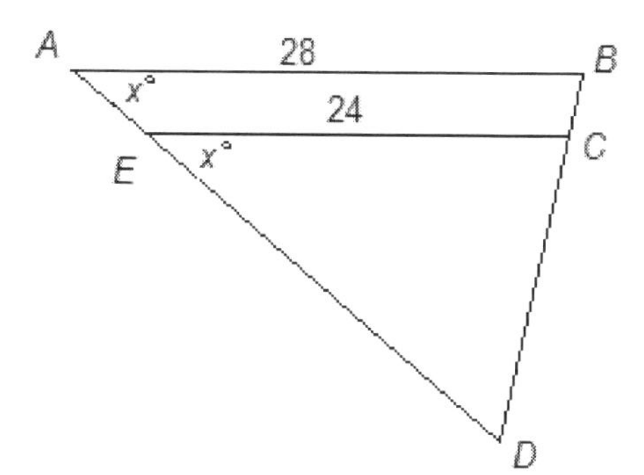


23. In the figure above, what is the value of $\frac{ED}{AD}$?

(A) $\frac{1}{7}$

(B) $\frac{1}{4}$

(C) $\frac{2}{5}$

(D) $\frac{1}{2}$

(E) $\frac{6}{7}$

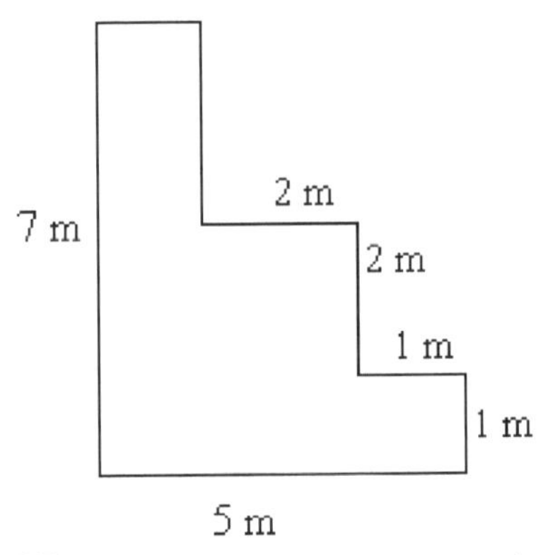


Note: Figure not drawn to scale.

24. What is the perimeter of the figure above?

    (A) 16 m
    (B) 18 m
    (C) 20 m
    (D) 22 m
    (E) 24 m

# LEVEL 1: PROBABILITY AND STATISTICS

25. A menu lists 2 appetizers, 5 meals, 4 drinks, and 3 desserts. A dinner consists of 1 of each of these 4 items. How many different dinners are possible from this menu?

    (A) 2
    (B) 4
    (C) 14
    (D) 72
    (E) 120

26. The average of $x$, $y$, $z$, and $w$ is 12 and the average of $z$ and $w$ is 7. What is the average of $x$ and $y$?

    (A) 14
    (B) 15
    (C) 16
    (D) 17
    (E) 18

3, 6, 7, 21, 27, 35, 42, 63, 70

27. A number is to be selected at random from the list above. What is the probability that the number selected will be a multiple of both 3 and 7?

(A) $\frac{1}{9}$

(B) $\frac{2}{9}$

(C) $\frac{1}{3}$

(D) $\frac{4}{9}$

(E) $\frac{5}{9}$

28. Exactly 4 musicians try out to play 4 different instruments for a particular performance. If each musician can play each of the 4 instruments, in how many ways can the 4 musicians be assigned to the 4 instruments?

(A) 1
(B) 4
(C) 12
(D) 24
(E) 256

# LEVEL 1: TRIGONOMETRY

29. Let $x = \cos\theta$ and $y = \sin\theta$ for any real value $\theta$. Then $x^2 - y^2 =$

(A) $-1$
(B) 0
(C) 1
(D) 2
(E) It cannot be determined from the information given

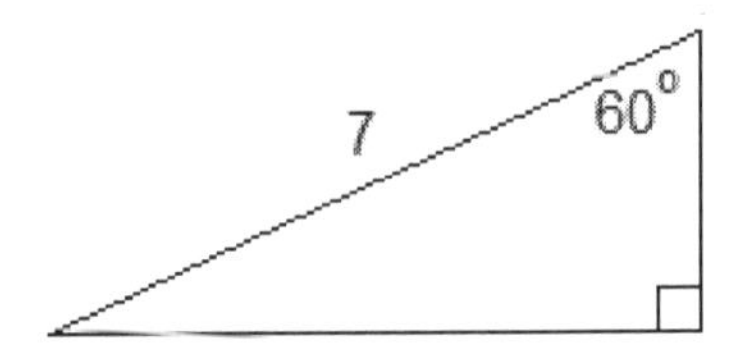


30. The figure above shows a right triangle whose hypotenuse is 7 feet long. How many feet long is the longer leg of this triangle?

    (A) 3.5
    (B) 14
    (C) $\frac{7\sqrt{3}}{2}$
    (D) $\frac{7\sqrt{3}}{6}$
    (E) $\frac{14\sqrt{3}}{3}$

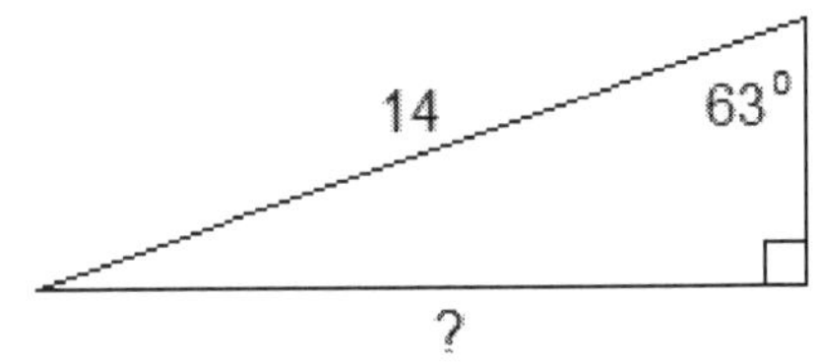


31. As shown above, a 14-foot ramp forms an angle of 63° with the vertical wall it is leaning against. Which of the following is an expression for the horizontal length, in feet, of the ramp?

    (A) 14 cos 63°
    (B) 14 sin 63°
    (C) 14 tan 63°
    (D) 14 sin 27°
    (E) 14 tan 27°

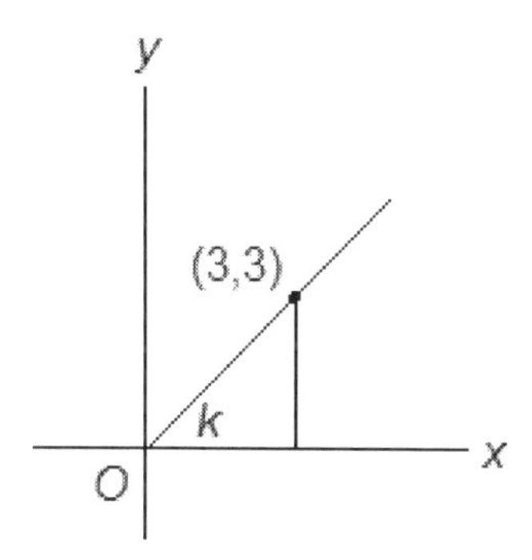


32. In the figure above, tan $k$ = ?

    (A) $3\sqrt{2}$

    (B) $\frac{\sqrt{2}}{2}$

    (C) $\frac{\sqrt{3}}{2}$

    (D) 1

    (E) $\frac{1}{3}$

# LEVEL 2: NUMBER THEORY

33. Alex plans to drive 550 miles from New York to Virginia, driving an average of 50 miles per hour. How many miles per hour faster must he average, while driving, to reduce his total driving time by 1 hour?

    (A) 20
    (B) 18
    (C) 15
    (D) 8
    (E) 5

S T A C K S T A C...

34. In the pattern above, the first letter is S and the letters S, T, A, C, and K repeat continually in that order. What is the 97th letter in the pattern?

(A) S
(B) T
(C) A
(D) C
(E) K

35. The ratio of the number of elephants to the number of zebras in a zoo is 3 to 5. What percent of the animals in the zoo are zebras?

(A) 12.5%
(B) 37.5%
(C) 60%
(D) 62.5%
(E) 70%

36. For integers $a$ and $b$, let $a \star b$ be the largest integer that divides both $a$ and $b$. The positive difference between $100 \star 150$ and $80 \star 120$ is

(A) 25
(B) 20
(C) 15
(D) 10
(E) 5

# LEVEL 2: ALGEBRA AND FUNCTIONS

37. Which of the following numbers is NOT contained in the domain of the function $h$ if $h(x) = \frac{x^2-2}{2x-1} + \frac{3}{x+2}$?

(A) $-1$
(B) 0
(C) $\frac{1}{2}$
(D) $\frac{3}{2}$
(E) 2

38. If $f(x, y) = \frac{x^2 - y^2}{x - y}$, what is $f(1.5, 2.5)$ ?

(A) $-1$
(B) 0
(C) 2
(D) 3
(E) 4

$$ax + by = 17$$
$$ax + (b + 1)y = 26$$

39. Based on the equations above, which of the following must be true?

(A) $x = 13.5$
(B) $x = 18$
(C) $y = 4.5$
(D) $y = 9$
(E) $x - y = 4.5$

40. If $y$ varies directly as $x$ and $y = 5$ when $x = 8$, then what is $y$ when $x = 24$?

(A) 3
(B) 5
(C) 15
(D) 20
(E) 24

| $x$ | $p(x)$ | $q(x)$ | $r(x)$ |
|---|---|---|---|
| -2 | -3 | 4 | -3 |
| -1 | 2 | 1 | 2 |
| 0 | 5 | -1 | -6 |
| 1 | -7 | 0 | -5 |

41. The functions $p$, $q$ and $r$ are defined for all values of $x$, and certain values of those functions are given in the table above. What is the value of $p(-2) + q(0) - r(1)$?

(A) $-2$
(B) $-1$
(C) 0
(D) 1
(E) 2

42. What are all values of $x$ for which $|x+3| \geq 4$?

(A) $x \leq 1$
(B) $x \leq -7$ or $x \geq 1$
(C) $-7 \leq x \leq 1$
(D) $-1 \leq x \leq 7$
(E) $x \geq 1$

43. If $15^k = 3^m \cdot 5^m$, then $k =$

(A) 0
(B) $\frac{m}{2}$
(C) $m$
(D) $2m$
(E) $11m$

44. For all $x \neq 0$, $\frac{2}{(\frac{3}{x^5})} =$

(A) $6x^5$
(B) $\frac{1}{6x^5}$
(C) $2\frac{3}{x^5}$
(D) $\frac{3x^5}{16}$
(E) $\frac{2x^5}{3}$

45. If $x \neq 0$ and $x$ is directly proportional to $y$, which of the following is inversely proportional to $\frac{1}{y^2}$?

(A) $x^2$
(B) $x$
(C) $\frac{1}{x}$
(D) $\frac{1}{x^2}$
(E) $-\frac{1}{x^2}$

46. The operation ⊠ is defined as $a \boxtimes b = \sqrt{a^3 + b^5}$ where $a$ and $b$ are positive real numbers. What is the value of $2 \boxtimes 1$ ?

(A) 0
(B) 1
(C) 2
(D) 3
(E) 4

# Level 2: Geometry

47. At what point does the graph of $5x - 2y - 20 = 0$ intersect the $x$-axis?

(A) $(-4,0)$
(B) $(-2,0)$
(C) $(0,0)$
(D) $(2,0)$
(E) $(4,0)$

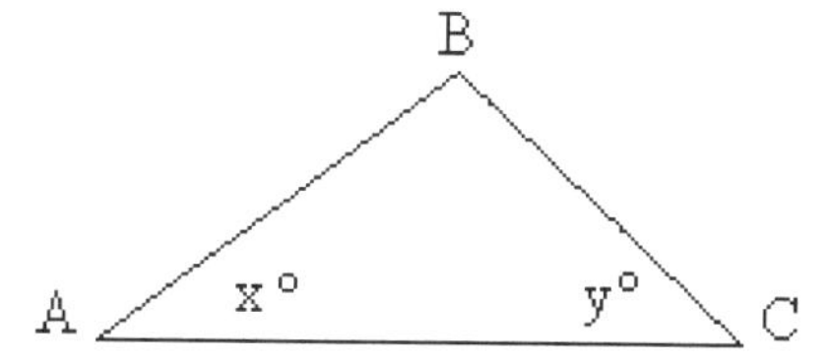

Note: Figure not drawn to scale.

48. In $\Delta ABC$ above, if $x < y$, which of the following must be true?

(A) $AB < BC$
(B) $AB < AC$
(C) $AB = BC$
(D) $AB > BC$
(E) $AB > AC$

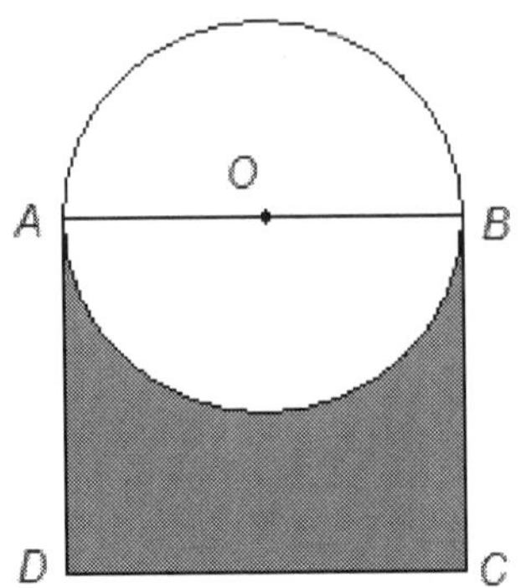


49. In the figure above, $AB$ is a diameter of the circle with center $O$ and $ABCD$ is a square. What is the area of the shaded region if the radius of the circle is 5?

(A) $25(4 - \frac{\pi}{2})$
(B) $25(2 - \frac{\pi}{2})$
(C) $\pi(4 - \pi)$
(D) $\pi(2 - \pi)$
(E) $\pi(1 - \pi)$

50. If line $m$ is perpendicular to segment $PQ$ at point $R$ and $PR = RQ$, how many points on line $m$ are equidistant from point $P$ and point $Q$?

(A) One
(B) Two
(C) Three
(D) Four
(E) More than four

51. In the $xy$-coordinate plane, line $n$ passes through the points (0,5) and (-2,0). If line $m$ is perpendicular to line $n$, what is the slope of line $m$?

(A) $-\frac{5}{2}$
(B) $-\frac{2}{5}$
(C) 1
(D) $\frac{2}{5}$
(E) $\frac{5}{2}$

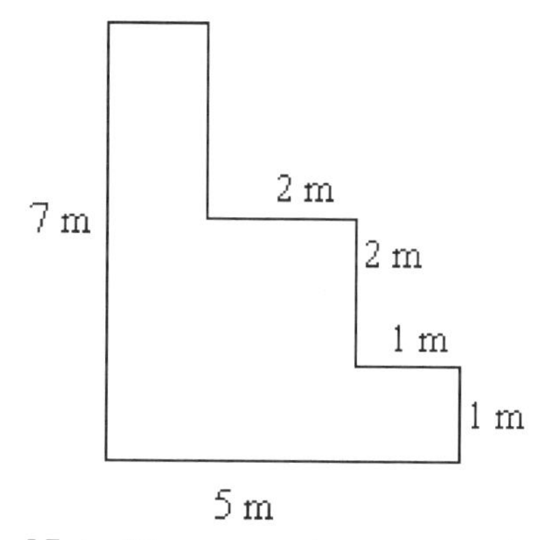


Note: Figure not drawn to scale.

52. What is the area of the figure above?

    (A) 15 $m^2$
    (B) 17 $m^2$
    (C) 19 $m^2$
    (D) 21 $m^2$
    (E) 23 $m^2$

53. Line $k$ contains the point (4,0) and has slope 5. Which of the following points is on line $k$?

    (A) (1, 5)
    (B) (3, 5)
    (C) (5, 5)
    (D) (7, 5)
    (E) (9, 5)

54. Point $O$ lies in plane $P$. How many circles are there in plane $P$ that have center $O$ and an area of $100\pi$ centimeters?

    (A) None
    (B) One
    (C) Two
    (D) Three
    (E) More than three

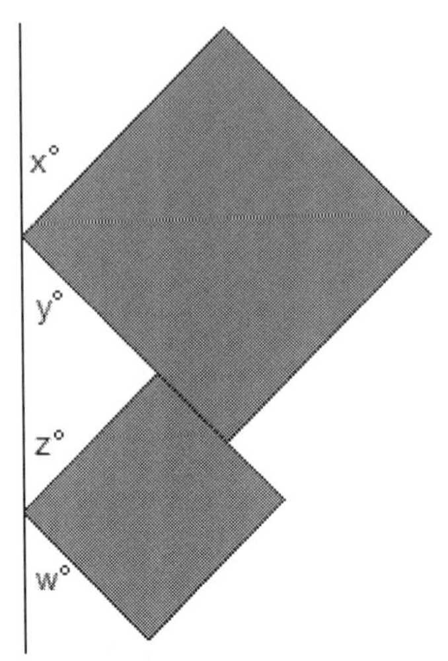


Note: Figure not drawn to scale.

55. In the figure above, the two shaded regions are squares. Which of the following must be true?

    (A) $x = y$
    (B) $x = w$
    (C) $y = z$
    (D) $y = w$
    (E) $z = w$

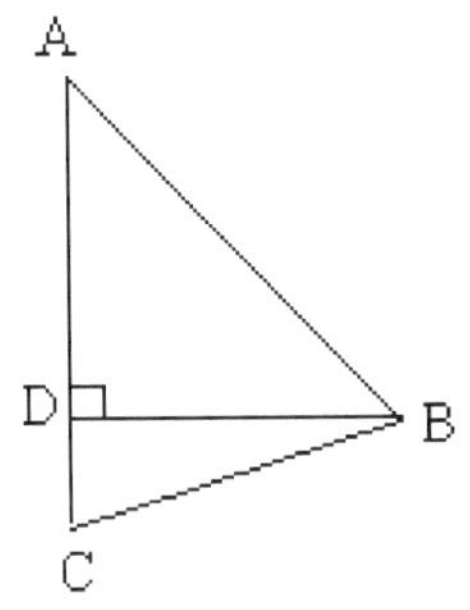


Note: Figure not drawn to scale.

56. Triangle $ABC$ has the same area as a rectangle with sides of lengths 5 and 7. If the length of $AC$ is 10, what is the length of $BD$?

    (A) 4
    (B) 5
    (C) 6
    (D) 7
    (E) 8

# Level 2: Probability and Statistics

57. The average (arithmetic mean) of $z$, 2, 16, and 21 is $z$. What is the value of $z$?

    (A) 12
    (B) 13
    (C) 14
    (D) 15
    (E) 16

58. In Bakerfield, 60 of the residents who own at least one cat also play the piano. If 200 residents of Bakerfield do not own a cat, and Bakerfield has 400 residents, how many residents own at least one cat but do not play the piano?

    (A) 100
    (B) 120
    (C) 140
    (D) 160
    (E) 180

59. If two six-sided dice are rolled, what is the probability that the sum of the two numbers will be 9?

    (A) $\frac{1}{9}$
    (B) $\frac{1}{8}$
    (C) $\frac{1}{6}$
    (D) $\frac{1}{4}$
    (E) $\frac{1}{3}$

60. From a group of 7 people, in how many ways can 4 be chosen?

    (A) 4
    (B) 7
    (C) 28
    (D) 35
    (E) 210

# LEVEL 2: TRIGONOMETRY

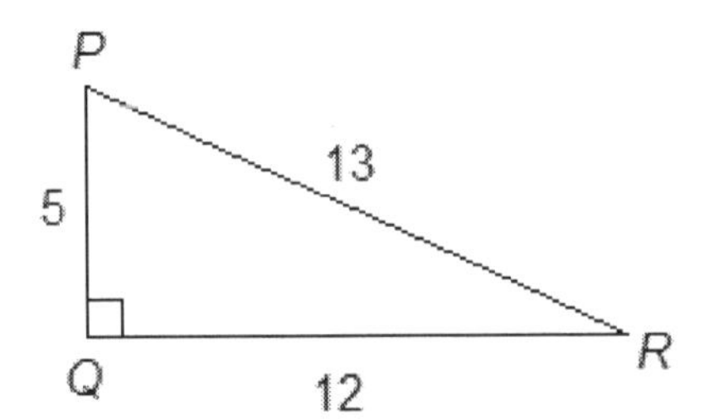


61. For $\angle P$ in $\Delta PQR$ above, which of the following trigonometric expressions has value $\frac{5}{13}$?

(A) $\tan P$
(B) $\cos P$
(C) $\sin P$
(D) $\cos R$
(E) $\tan R$

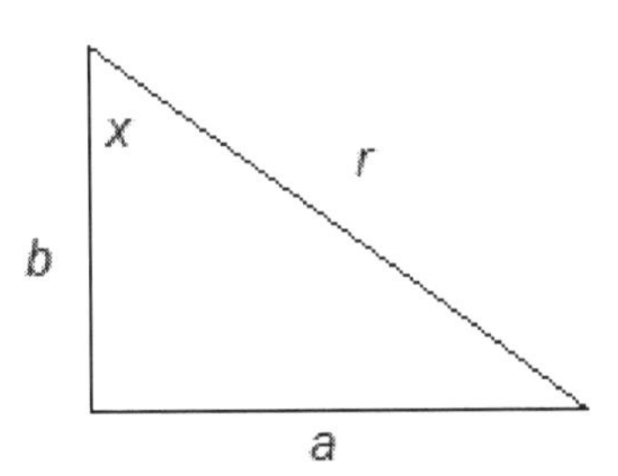


62. In the figure above, $r\sin x =$

(A) $a$
(B) $b$
(C) $r$
(D) $a + b$
(E) $ab$

63. For any acute angle with measure $A$, $\cos(90° - A) =$

(A) $\sin A$
(B) $\cos A$
(C) $\tan A$
(D) $\csc A$
(E) $\sec A$

64. The vertex of $\angle P$ is the origin of the standard $(x, y)$ coordinate plane. One ray of $\angle P$ is the positive $x$-axis. The other ray, $\overrightarrow{PQ}$, is positioned so that $\cos A < 0$ and $\tan A > 0$. In which quadrant, if it can be determined, is point $Q$ ?

(A) Quadrant I
(B) Quadrant II
(C) Quadrant III
(D) Quadrant IV
(E) Cannot be determined from the given information

# LEVEL 3: NUMBER THEORY

65. Which of the following has the greatest value?

(A) 500,000,000,000
(B) $50^{1000}$
(C) $5000^{100}$
(D) $(50 \cdot 50^{100})^{100}$
(E) $(500 \cdot 50)^{1000}$

66. A piece of paper is cut into two pieces. A second piece of paper is then cut into four pieces, so that there are a total of six pieces of paper. A third piece of paper is then cut into six pieces, so that there are a total of twelve pieces of paper. If this process continues, how many pieces of paper are there after the $n$th piece of paper is finished being cut?

(A) $n(n + 1)$
(B) $n(n - 1)$
(C) $\frac{n(n+1)}{2}$
(D) $2n^2$
(E) $2n^2 + 1$

67. If $\frac{3}{5}$ is $\frac{2}{7}$ of $\frac{14}{3}$ of a certain number, what is the number?

(A) $\frac{2}{5}$
(B) $\frac{9}{20}$
(C) $\frac{11}{20}$
(D) $\frac{3}{5}$
(E) $\frac{3}{4}$

68. In a geometric sequence, the third term is 4 and the fifth term is 9. What is the seventh term in the sequence?

(A) 243
(B) 81
(C) 27
(D) $\frac{27}{2}$
(E) $\frac{81}{4}$

## LEVEL 3: ALGEBRA AND FUNCTIONS

69. A printing company has daily operating expenses of $12,000. On Monday the company had 1000 customers and made a profit of $5000. On Tuesday the company had 1500 customers. Assuming that each customer paid the same amount of money, what profit did the company make on Tuesday?

(A) $6000
(B) $7000
(C) $7500
(D) $12,000
(E) $13,500

70. What are all values of $x$ for which $16 - x^2 \geq 1 - 2x$ ?

(A) $x \leq -3$
(B) $x \geq 5$
(C) $-3 \leq x \leq 5$
(D) $-5 \leq x \leq 3$
(E) $x \leq -3$ or $x \geq 5$

-3 -2 0 2 3

71. The number line graph above could be the graph of all values of $x$ for which

(A) $2 \leq x^2 \leq 3$
(B) $4 \leq x^2 \leq 9$
(C) $2 \leq x \leq 3$
(D) $4 \leq x \leq 9$
(E) $-3 \leq x \leq 2$

72. If $(x-2)g(x) = x^3 - 5x^2 + 10x - 8$ where $g(x)$ is a polynomial in $x$, then $g(x) =$

(A) $x + 4$
(B) $x^2 + 4$
(C) $x^2 + 4x$
(D) $x^2 - 3x + 4$
(E) $x^2 + 3x + 4$

73. If $rs = 4, st = 7, rt = 63$, and $r > 0$, then $rst =$

(A) 35
(B) 40
(C) 42
(D) 60
(E) 120

74. $\frac{5x(yz+z)-5xz}{-xyz} =$

(A) $-5$
(B) $-4$
(C) $-3$
(D) $-2$
(E) $-1$

75. If $k(x) = \frac{x^2-4}{x-2}$ and $h(x) = \frac{3x}{8}$, then $h(k(\pi)) =$

(A) 2.56
(B) 1.23
(C) 0
(D) $-2.56$
(E) $-3.12$

76. How many roots does the function $f(x) = x^3 + x^2 - 2x$ have?

(A) One
(B) Two
(C) Three
(D) Four
(E) More than four

77. For all real numbers $a$ and $b$, $|b - a| - |a - b| =$

(A) $2(a - b)$
(B) $2b - 2a$
(C) 0
(D) $2b + 2a$
(E) $2a + b$

78. If $x$ varies inversely as $y$, and $x = 2$ when $y = 6$, then what is the value of $y$ when $x = k$ ?

(A) $12k^2$
(B) $6k^2$
(C) $12k$
(D) $\frac{12}{k}$
(E) $\frac{4}{k}$

# LEVEL 3: GEOMETRY

79. Line $k$ has a negative slope and a positive $y$-intercept. Line $j$ is parallel to line $k$ and has a negative $y$-intercept. The $x$-intercept of $j$ must be

(A) positive and less than the $x$-intercept of $k$
(B) negative and less than the $x$-intercept of $k$
(C) positive and greater than the $x$-intercept of $k$
(D) negative and greater than the $x$-intercept of $k$
(E) zero

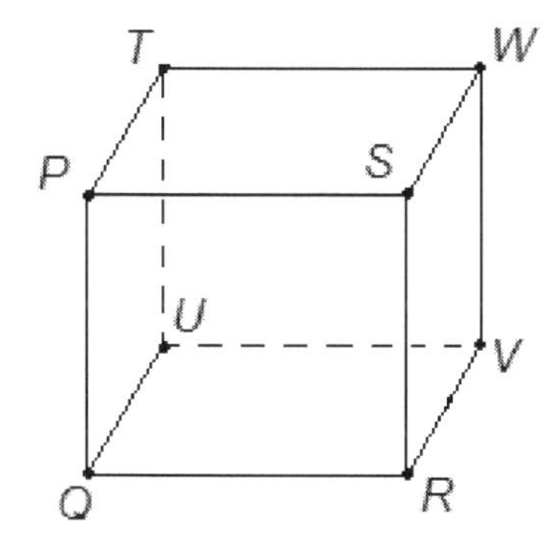


80. The figure above is a cube. Which of the given points is in the plane determined by vertices $S$, $R$, and $T$?

    (A) $P$
    (B) $Q$
    (C) $U$
    (D) $V$
    (E) $W$

81. The points $O(0,0)$, $A(0,-5)$, $B(-2,3)$, and $C(6,1)$ are plotted in the $xy$-plane. Which of the following two segments have the same length?

    (A) $OA$ and $OB$
    (B) $OA$ and $OC$
    (C) $OB$ and $OC$
    (D) $AB$ and $AC$
    (E) $AB$ and $BC$

82. In the $xy$-plane, the point $(-3,5)$ is the midpoint of the line segment with endpoints $(1,y)$ and $(x,4)$. What is the value of $y-x$?

    (A) $-13$
    (B) $-4$
    (C) 1
    (D) 4
    (E) 13

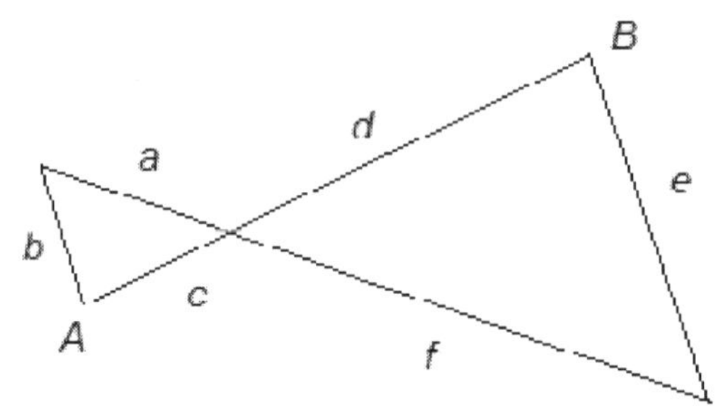


83. In the figure above, the two triangles are similar with both $m\angle A = 87°$ and $m\angle B = 87°$. If $\frac{a}{c} = 4$, then $\frac{d}{f} =$

(A) 8
(B) 4
(C) 2
(D) $\frac{1}{2}$
(E) $\frac{1}{4}$

84. If the base radius of cone $S$ is one-half as long as the base radius of cone $T$, then the volume of cone $S$ is what fraction of the volume of cone $T$ ?

(A) $\frac{1}{2}$
(B) $\frac{1}{3}$
(C) $\frac{1}{4}$
(D) $\frac{1}{5}$
(E) $\frac{1}{8}$

85. In the rectangular coordinate system the point $P(a, b)$ is moved to the new point $Q(5a, 5b)$. What is the distance between points $P$ and $Q$?

(A) $a$
(B) $b$
(C) $\sqrt{a^2 + b^2}$
(D) $2\sqrt{a^2 + b^2}$
(E) $4\sqrt{a^2 + b^2}$

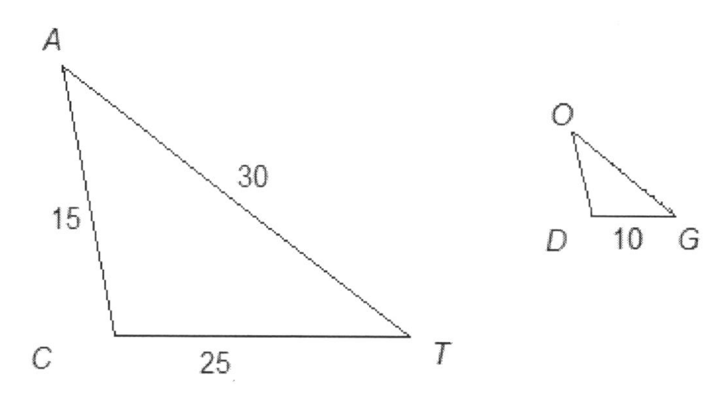


86. In the figure above, where $\Delta CAT \sim \Delta DOG$, lengths given are in inches. What is the perimeter, in inches, of $\Delta DOG$ ?

(Note: The symbol ~ means "is similar to.")

(A) 35
(B) 28
(C) 21
(D) 14
(E) 7

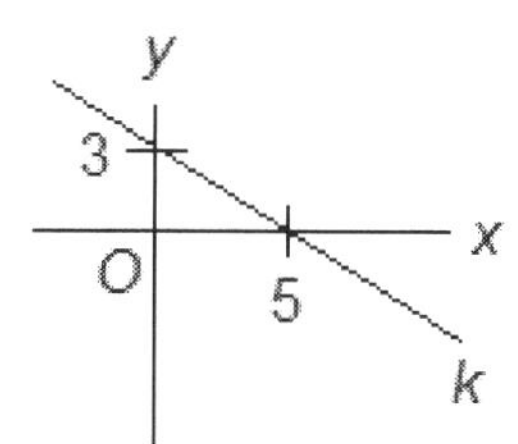


87. An equation of line $k$ in the figure above is

(A) $x = 5$

(B) $y = 3$

(C) $y = \frac{3}{5}x + 3$

(D) $y = -\frac{3}{5}x + 3$

(E) $y = \frac{5}{3x} + 3$

88. Lines $k$ and $n$ are distinct parallel lines. If line $n$ passes through the points (0,0) and (3,2), then line $k$ can pass through which of the following points?

(A) (–3,–2)
(B) (–1,–3)
(C) (6,4)
(D) (9,6)
(E) (15,10)

## LEVEL 3: PROBABILITY AND STATISTICS

89. In a small town, 30 families have cats, 50 families have dogs, and 25 families have neither cats nor dogs. If there are 90 families living in the town, how many families have both cats and dogs?

(A) 10
(B) 15
(C) 20
(D) 25
(E) 30

90. If two six-sided dice are rolled, what is the probability that the sum of the two numbers shown will be even?

(A) $\frac{1}{2}$
(B) $\frac{1}{3}$
(C) $\frac{1}{4}$
(D) $\frac{1}{6}$
(E) $\frac{1}{9}$

91. In how many different ways can a line of 7 people be formed?

(A) 7
(B) 42
(C) 49
(D) 5040
(E) 823,543

TEST GRADES OF STUDENTS IN MATH CLASS

| Test Grade | 75 | 82 | 87 | 93 | 100 |
|---|---|---|---|---|---|
| Number of students with that grade | 5 | 7 | 10 | 3 | 1 |

92. The test grades of the 26 students in a math class are shown in the chart above. What is the median test grade for the class?

    (A) 75
    (B) 82
    (C) 87
    (D) 90
    (E) 93

## LEVEL 3: TRIGONOMETRY

93. In triangle $PQR$ with right angle $Q$, the length of side $QR$ is 7 and the measure of $\angle PRQ$ is 43°. What is the length of side $PR$ ?

    (A) 0.1
    (B) 1.4
    (C) 7.5
    (D) 9.6
    (E) 10.2

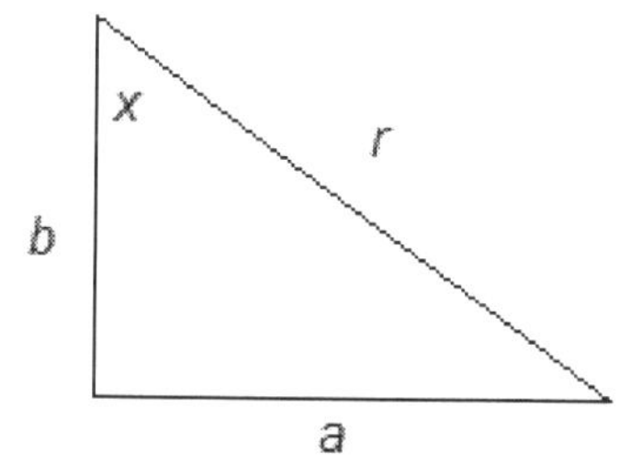


94. In the figure above, $\frac{b}{a}\tan x =$

    (A) 1
    (B) $-1$
    (C) $r$
    (D) $a + b$
    (E) $ab$

95. $(2 - \cos^2\theta - \sin^2\theta)^2$ is equal to

(A) 0
(B) 1
(C) 2
(D) 32
(E) 243

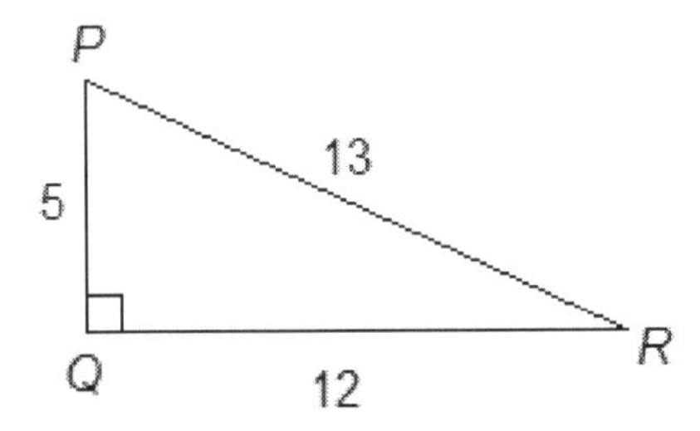


96. For $\angle R$ in $\Delta PQR$ above, which of the following trigonometric expressions has value $\frac{5}{12}$ ?

(A) $\tan R$
(B) $\cot R$
(C) $\sin R$
(D) $\csc R$
(E) $\sec R$

## LEVEL 4: NUMBER THEORY

97. If the positive integers, starting with 1, are written consecutively, what will be the 120th digit written?

(A) 0
(B) 2
(C) 4
(D) 6
(E) 8

98. In 2000, the population of Bakersfield was 8000. If the population of Bakersfield increases at the rate of 2.6% each year, what will the population of Bakersfield be at the end of 2020?

(A) 13,367
(B) 12,258
(C) 11,182
(D) 10,221
(E) 8,233

99. Two integers $m$ and $n$ satisfy the relation $m \vdots n$ if and only if $m = 5n^3 - 2$. If $a$, $b$, and $c$ satisfy the relations $a \vdots b$ and $b \vdots c$, what is $a$ in terms of $c$ ?

(A) $5(5c^3 - 2)^3 - 4$
(B) $5(5c^3 - 2)^3 - 2$
(C) $5(5c^3 - 2) - 2$
(D) $5c^3 - 4$
(E) $5c^3 - 2$

100. A drink is made by mixing juice with water. How many quarts of juice should be mixed with 3 quarts of water so that 72 percent of the drink is juice?

(A) 5.9
(B) 6.8
(C) 7.2
(D) 7.7
(E) 8.3

# LEVEL 4: ALGEBRA AND FUNCTIONS

101. In the equation $5x^2 - 6x + 3 = 0$, let $s$ be the sum of the roots, and let $p$ be the product of the roots. If $s = kp$, then $k =$

(A) −4
(B) −2
(C) 0
(D) 2
(E) 4

102. If $a$ and $b$ are in the domain of a function $g$ and $g(a) = g(b)$, which of the following must be true?

(A) $a = b$
(B) $g$ fails the vertical line test.
(C) there is a horizontal line that intersects the graph of $g$ at least twice.
(D) the graph of $g$ is a horizontal line.
(E) The point $(a, b)$ is on the graph of $g$.

103. If $f(x) = \sqrt{3 - x}$ and $g(x) = \frac{1}{x^2-1}$, what is the domain of $fg$ ?

(A) all real numbers $x$

(B) all $x$ such that $x \geq 3$

(C) all $x$ such that $x \neq 1$ and $x \leq 3$

(D) all $x$ such that $x \neq \pm 1$ and $x \leq 3$

(E) all $x$ such that $1 < x \leq 3$

104. If $x = \frac{2}{5}$ is a solution to the equation $3(x - 5)(10x - c) = 0$, what is the value of $c$ ?

(A) 4
(B) 3
(C) 2
(D) 1
(E) 0

105. What value does $\frac{x^2-3x-10}{5x+10}$ approach as $x$ approaches $-2$?

(A) $-\frac{7}{5}$
(B) $-1$
(C) $-\frac{5}{7}$
(D) $\frac{1}{5}$
(E) 5

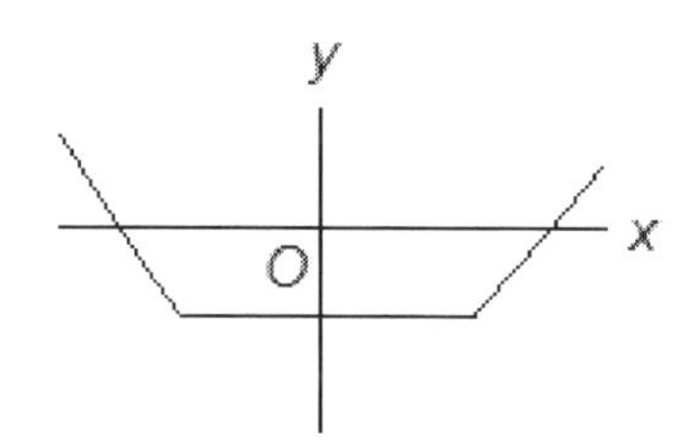


106. The graph of $y = h(x)$ is shown above. Which of the following could be the graph of $y = h(|x|)$ ?

(A)

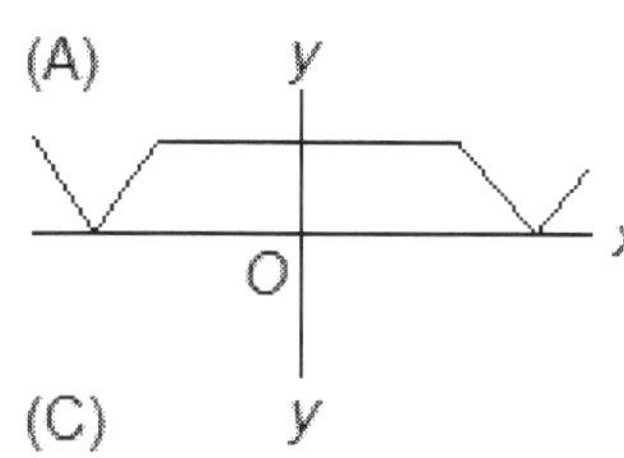


(B)

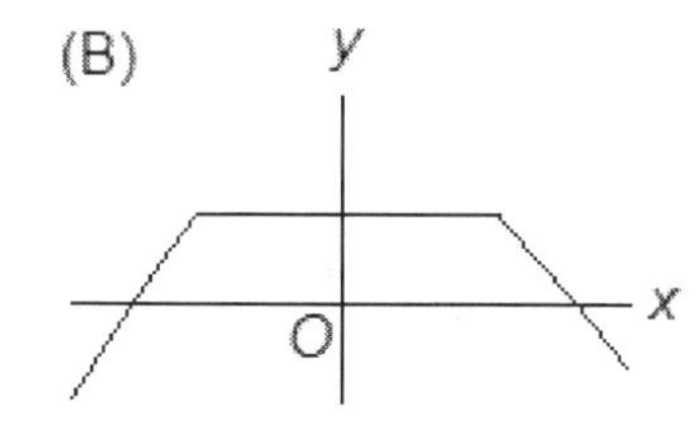


(C)

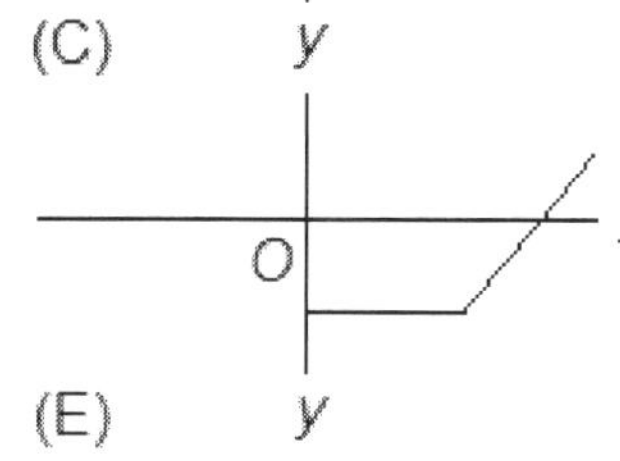


(D)

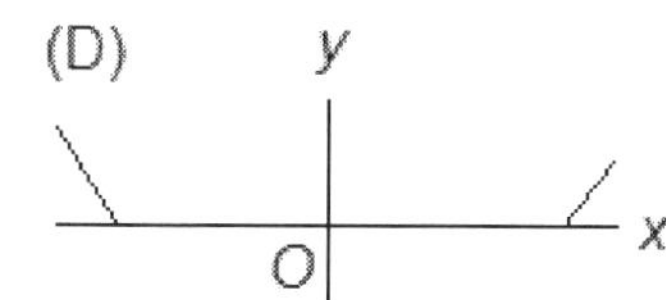


(E)

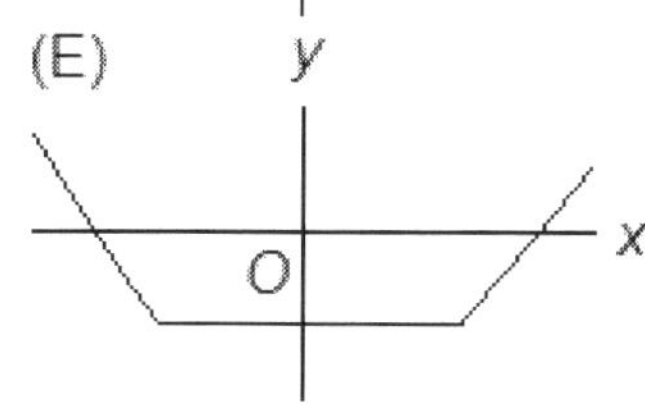


107. If $k(x) = \left(\sqrt{x} + 3\right)^3$, for all $x > 0$, then which of the following functions is equal to $f^{-1}(x)$ when restricted to $x > 27$ ?

(A) $\sqrt[3]{x}$
(B) $\sqrt[3]{x} - 3$
(C) $\left(\sqrt[3]{x} - 3\right)^2$
(D) $(x - 3)^2$
(E) $(x - 3)^3$

108. In the equation $x^2 - bx + c = 0$, $b$ and $c$ are integers. The solutions of this equation are 2 and 3. What is $b - c$ ?

(A) $-11$
(B) $-1$
(C) 1
(D) 5
(E) 11

109. If $c > 0$, $s^2 + t^2 = c$, and $st = c + 5$, what is$(s + t)^2$ in terms of $c$ ?

(A) $c + 5$
(B) $c + 10$
(C) $2c + 5$
(D) $2c + 10$
(E) $3c + 10$

110. When $x \neq 7$, $\frac{3x}{x^2-49} + \frac{3x}{7-x}$ is equivalent to:

(A) $\frac{-3x^2-21x}{x^2-49}$

(B) $\frac{-21x}{x^2-49}$

(C) $\frac{6-21x}{x^2-49}$

(D) $\frac{-3x^2-18x}{x^2-49}$

(E) $\frac{-3x^2}{x^2-49}$

# LEVEL 4: GEOMETRY

111. In the $xy$-plane, which of the following are the points of intersection of the line $x = 5$ with the circle centered at (4,2) with radius 4 ?

(A) $(5, -1.87)$, $(5, 5.87)$
(B) $(5, -2)$, $(5, 2)$
(C) $(5, -4)$, $(5, 4)$
(D) $(5, -3.16)$, $(5, 6.24)$
(E) $(5, -16)$, $(5, 16)$

112. Three right circular cylinders, each with a height of 2 feet, have base radii of 3, 6 and $x$, respectively. If the volume of the cylinder with base radius $x$ is the arithmetic mean of the volumes of the other two cylinders, then $x =$

(A) 4.50
(B) 4.74
(C) 5.0
(D) 22.5
(E) 25.0

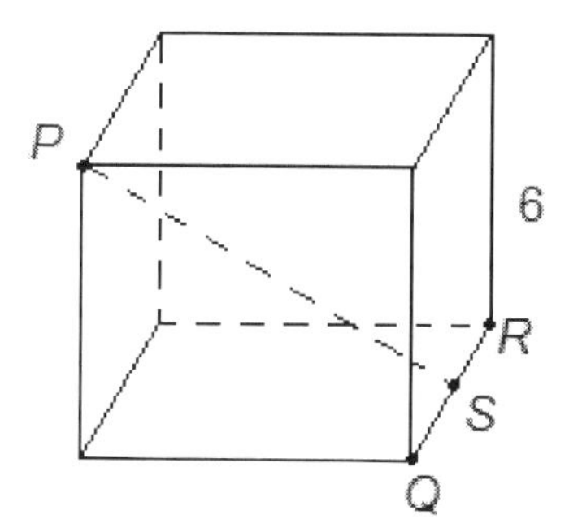


113. The cube shown above has a side of length 2. What is the distance from vertex $P$ to the midpoint $S$ of edge $QR$ ?

(A) 3
(B) $3\sqrt{3}$
(C) 9
(D) $6\sqrt{3}$
(E) 81

114. $AB$ is a diameter of circle $O$ and $C$ is a point on the circle distinct from $A$ and $B$. Triangle $ABC$ must be

(A) equilateral
(B) isosceles
(C) scalene
(D) right
(E) obtuse

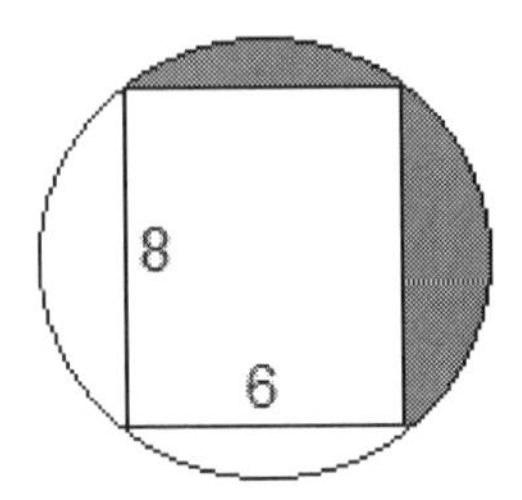


115. In the figure above, the rectangle with sides of length 6 and 8 is inscribed in the circle. What is the area of the shaded region?

(A) 15.3
(B) 18.7
(C) 30.5
(D) 48.2
(E) 61.1

116. Line $k$ is the perpendicular bisector of the line segment with endpoints (3,1) and (5,5). What is the slope of line $k$ ?

(A) $-2$
(B) $-\frac{1}{2}$
(C) 0
(D) $\frac{1}{2}$
(E) 2

117. What is the volume of a cube with surface area $S$ square meters?

(A) $S\sqrt{S}$
(B) $S\sqrt{6}$
(C) $S\sqrt{6S}$
(D) $\frac{S\sqrt{6S}}{6}$
(E) $\left(\sqrt{\frac{S}{6}}\right)^3$

118. Stones are placed inside a circular region so that there are 6 stones per square meter. If the circumference of the region is 15 meters, how many stones are needed?

(A) 84
(B) 90
(C) 96
(D) 102
(E) 108

119. A triangle with vertices $A(-2,1)$, $B(3,2)$, and $C(1,3)$ is reflected across the line $y = -1$. What will be the coordinates of the reflection of point $B$ ?

(A) $(3,0)$
(B) $(3,-1)$
(C) $(3,-2)$
(D) $(3,-3)$
(E) $(3,-4)$

120. In quadrilateral $ABCD$, $100 < m\angle A + m\angle B < 200$. Which of the following describes all possible values of $m\angle C + m\angle D$ ?

(A) $360 < m\angle C + m\angle D < 460$
(B) $160 < m\angle C + m\angle D < 260$
(C) $120 < m\angle C + m\angle D < 160$
(D) $80 < m\angle C + m\angle D < 120$
(E) $0 < m\angle C + m\angle D < 80$

# LEVEL 4: PROBABILITY AND STATISTICS

121. In a list of positive integers, 80% of the integers are greater than or equal to 150. Which of the following must be greater than or equal to 150?

(A) The mode of the integers in the list
(B) The median of the integers in the list
(C) The mean of the integers in the list
(D) The positive difference between the greatest integer and the least integer in the list
(E) The standard deviation of the integers in the list

122. If a 3-digit positive integer is chosen at random, what is the probability that an integer is chosen that contains only nonzero digits?

(A) 0.99
(B) 0.81
(C) 0.64
(D) 0.50
(E) 0.01

123. 30 people are asked how many siblings they have. 2 of them have no siblings, 5 have 1 sibling, 12 have 2 siblings, 6 have 3 siblings, and 5 have 4 siblings. If one of these people were chosen at random, what is the probability that this person has 2 or more siblings?

(A) 0.43
(B) 0.70
(C) 0.77
(D) 0.84
(E) 0.90

124. At a gathering, each of the seven people in attendance shakes hands with each of the other six people exactly three times. How many handshakes take place?

(A) 21
(B) 36
(C) 42
(D) 63
(E) 126

## LEVEL 4: TRIGONOMETRY

125. To the nearest tenth of a degree, what is the measure of the smallest angle in a right triangle with sides 5, 12, and 13 ?

(A) 22.6°
(B) 32.1°
(C) 41.7°
(D) 52.0°
(E) 55.8°

126. $\frac{\sin x}{\cos x \tan x} =$

(A) $\cot x$
(B) $\csc x$
(C) $\sec x$
(D) $\cos x$
(E) None of these

127. If $\angle P$ is an acute angle and $\frac{\cos^2 P}{\sin^2 P} = 5.72$, what is the value of $\tan P$ ?

(A) 2.40
(B) 2.02
(C) 1.23
(D) 0.75
(E) 0.42

128. $(1 - \cos x)(1 + \cos x) =$

(A) $\sin x$
(B) $\cos x$
(C) $\sin^2 x$
(D) $\cos^2 x$
(E) $\sec x \tan x$

## LEVEL 5: NUMBER THEORY

129. A Fibonacci-type sequence can be defined recursively as $a_n = a_{n-1} + a_{n-2}$ for $n > 2$, and where $a_1$ and $a_2$ are given real numbers. What is the 8th term of the Fibonacci-type sequence where $a_1 = 2$ and $a_2 = 3$?

(A) 21
(B) 34
(C) 55
(D) 90
(E) 146

If Stanley gets a 90 on his last test, then he will get an A.

130. Which of the following CANNOT be inferred from the statement above?

(A) If Stanley gets an A, then he received a 90 on his last test.
(B) If Stanley did not get an A, then he did not receive a 90 on his last test.
(C) A necessary condition for Stanley to get a 90 on his last test is that he gets an A.
(D) In order for Stanley to get an A, it is sufficient that he gets a 90 on his last test.
(E) Stanley gets a 90 on his last test implies that he will get an A.

131. If $\log_4 x = k$, then $\log_2 x =$

(A) $\frac{k}{2}$

(B) $2k$

(C) $4k$

(D) $k^2$

(E) $2^k$

132. If $i^2 = -1$, then what is the value of $i^{73}$ ?

(A) 0
(B) 1
(C) $-1$
(D) $i$
(E) $-i$

# LEVEL 5: ALGEBRA AND FUNCTIONS

133. What is the range of the function defined by $\frac{1}{x^2} - 2$ ?

(A) All real numbers
(B) All real numbers except 0
(C) All real numbers except $-2$
(D) All real numbers greater than 0
(E) All real numbers greater than $-2$

134. The graph of the rational function $r$ where $r(x) = \frac{x^2-1}{x^2-4}$ has asymptotes $x = a$, $x = b$, and $y = c$. What is the value of $a + b + c$?

(A) $-\frac{1}{4}$

(B) $\frac{1}{4}$

(C) 1

(D) 3

(E) 5

135. If $x$ varies inversely as $y^2$, and $x$ is 3 when $y$ is 5, then what is $x$ when $y$ is 3?

(A) 3
(B) 5
(C) 8.3
(D) 16.6
(E) 25

136. If $g(x, y) = g(-x, -y)$ for all real numbers $x$ and $y$, then $g$ can be which of the following functions ?

(A) $x + y$
(B) $x^2 + y$
(C) $x^3 + y^2$
(D) $\frac{x+y}{x-y}$
(E) $\frac{x^2+y^3}{x^2-y^2}$

137. $\frac{x^4+x^3+x^2}{x^7+x^6+x^5} =$

(A) $x^3$

(B) $x$

(C) $x^{-3}$

(D) $3x$

(E) 3

138. If $ab^2c = 6$ and $a^2bc^3 = 27$, what is the value of $\frac{ac^2}{b}$ ?

(A) 0.2
(B) 3.0
(C) 4.5
(D) 5.4
(E) 6.0

139. Let $x \therefore y$ be defined as the sum of all integers between $x$ and $y$. For example, $1 \therefore 4 = 2 + 3 = 5$. What is the value of

$$(60 \therefore 900) - (63 \therefore 898)\ ?$$

(A) 1982
(B) 1983
(C) 1984
(D) 1985
(E) 1986

140. Suppose the graph of $g(x) = -2x^3$ is translated 2 units down and 3 units left. If the resulting graph represents $G(x)$, what is the value of $G(-.5)$ ?

(A) –33.25
(B) –28.75
(C) –17.25
(D) –11.5
(E) –2.25

141. The equation $3^{x^2-2x+8} = 243$ has two solutions. Let $a$ be the sum of these solutions and let $b$ be the product of these solutions. What is $a - b$ ?

(A) –2
(B) –1
(C) 0
(D) 1
(E) 2

142. What is the smallest possible value for the product of 2 real numbers that differ by 7?

(A) –49
(B) –12.25
(C) –7.25
(D) 0
(E) 12.25

# LEVEL 5: GEOMETRY

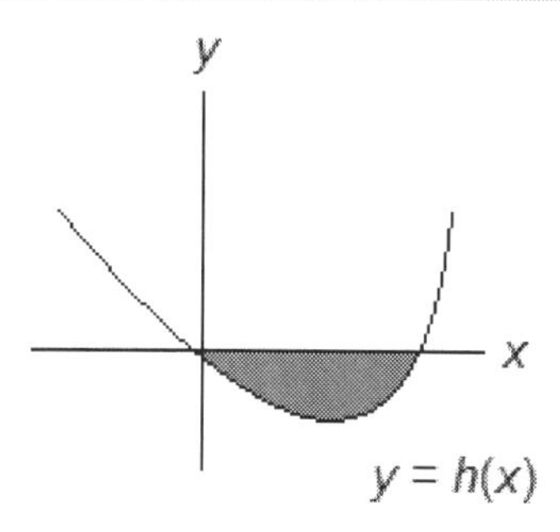


143. In the figure above, the area of the shaded region bounded by the graph of $y = h(x)$ and the $x$-axis is 5. The area of the region bounded by the graph of $y = -h(x + 4)$ and the $x$-axis is

(A) 1
(B) 2
(C) 2.5
(D) 5
(E) 9

144. If the measure of one angle of a rhombus is 30°, then the ratio of the length of its shorter diagonal to its longer diagonal is

(A) $\sqrt{2}$
(B) $\frac{2\sqrt{3}}{3}$
(C) $\frac{\sqrt{2}}{2}$
(D) $\frac{\sqrt{3}}{3}$
(E) $\frac{1}{2}$

145. A sphere with a radius of 10 is centered at the origin. Which of the following points is NOT inside the sphere?

(A) $(-5,5,-5)$
(B) $(4,-5,6)$
(C) $(1,7,-7)$
(D) $(-4,-4,8)$
(E) $(1,-8,-6)$

146. The set of points $(a,b,c)$ such that

$$(a-3)^2+(b-2)^2+(c+1)^2=-1$$

is

(A) a sphere
(B) a circle
(C) a parabola
(D) a point
(E) empty

147. What is the $x$-coordinate of the vertex of the parabola whose equation is $y=x^2-6x+3$ ?

(A) 0
(B) 1
(C) 2
(D) 3
(E) 4

148. A cone is inscribed in a cube of surface area 24 in such a way that its base touches four edges of the cube. What is the volume of the space enclosed by the cube, but NOT by the cone?

(A) 1.6
(B) 2.1
(C) 4.0
(D) 5.9
(E) 7.3

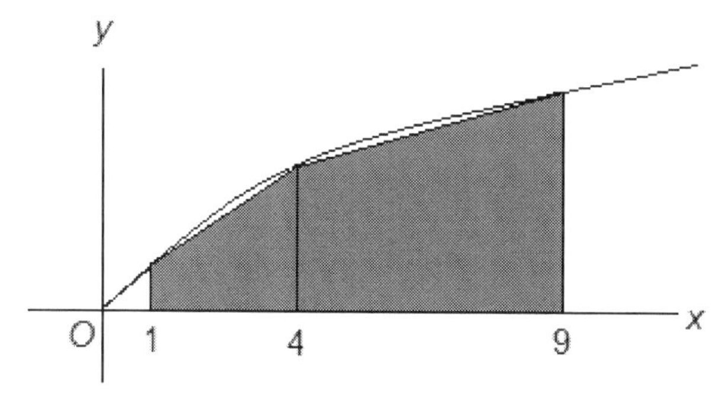


Note: Figure not drawn to scale.

149. The figure above shows a portion of the graph of $y = \sqrt{x}$. What is the sum of the areas of the two inscribed trapezoids shown?

(A) 11
(B) 17
(C) 23
(D) 34
(E) 38

150. A right circular cylinder has a base radius of 5 and a height of 4. If $P$ and $Q$ are two points on the surface of the cylinder, what is the maximum distance between $P$ and $Q$?

(A) 4.79
(B) 6.40
(C) 10.77
(D) 10.92
(E) 11.45

151. A sphere with volume 64 cubic inches is inscribed in a cube. What is the length of the long diagonal of the cube?

(A) 7.6
(B) 8.1
(C) 8.6
(D) 9.4
(E) 9.9

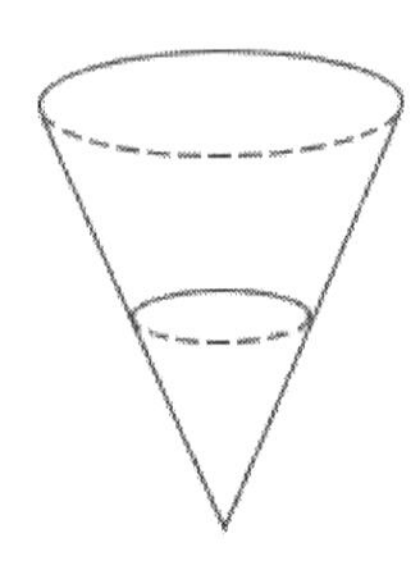

152. The right circular cone pictured above is cut horizontally at the midpoint of its height. If the smaller piece has a height of 3 and a volume of $9\pi$, what is the volume of the larger piece?

(A) $12\pi$
(B) $21\pi$
(C) $63\pi$
(D) $216\pi$
(E) $284\pi$

## LEVEL 5: PROBABILITY AND STATISTICS

153. A five digit number is randomly generated using each of the digits 1, 2, 3, 4, and 5 exactly once. What is the probability that the digits 2 and 3 are not next to each other?

(A) 0.2
(B) 0.3
(C) 0.4
(D) 0.5
(E) 0.6

154. A pile of books consists of 3 red books, 7 yellow books, and 5 blue books. If 2 books are selected at random from the pile, what is the probability that neither book is yellow?

(A) 0.73
(B) 0.61
(C) 0.50
(D) 0.33
(E) 0.27

155. A group of students take a test and the average score is 90. One more student takes the test and receives a score of 81 decreasing the average score of the group to 87. How many students were in the initial group?

(A) 2
(B) 3
(C) 4
(D) 5
(E) 6

156. A positive integer is called a palindrome if it reads the same forward as it does backward. For example, 1661 is a palindrome. If $n$ is a positive integer, how many palindromes are there of length $2n$?

(A) $n$
(B) $10^{n-1}$
(C) $10^n$
(D) $9 \cdot 10^{n-1}$
(E) $9 \cdot 10^n$

# LEVEL 5: TRIGONOMETRY

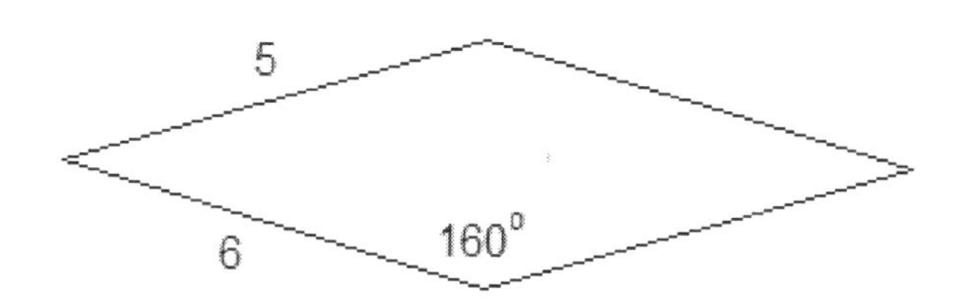


157. What is the area of the parallelogram shown in the figure above?

(A) 9.23
(B) 9.42
(C) 9.55
(D) 10.02
(E) 10.26

158. A 7 foot ladder is leaning against a wall such that the angle relative to the level ground is 70°. Which of the following expressions involving cosine gives the distance, in feet, from the base of the ladder to the wall?

(A) $\frac{7}{\cos 70°}$

(B) $\frac{\cos 70°}{7}$

(C) $\frac{1}{7\cos 70°}$

(D) $7\cos 70°$

(E) $\cos(7 \cdot 70°)$

159. For $0 < x < \frac{\pi}{2}$, the expression $\frac{\sin x}{\sqrt{1-\sin^2 x}} - \frac{\sqrt{1-\sin^2 x}}{\cos x}$ is equivalent to

(A) $\tan x$
(B) $\cot x$
(C) $\tan x - 1$
(D) $1 - \tan x$
(E) $1 - \cot x$

160. A ladder rests against the side of a wall and reaches a point that is 25 meters above the ground. The angle formed by the ladder and the ground is 58°. A point on the ladder is 3 meters from the wall. What is the vertical distance, in meters, from this point on the ladder to the ground?

(A) $22\tan 58°$
(B) $22\cos 58°$
(C) $25 - 3\sin 58°$
(D) $25 - 3\cos 58°$
(E) $25 - 3\tan 58°$

# Answers to Supplemental Problems

## Level 1: Number Theory

1. C
2. E
3. B
4. B

## Level 1: Algebra and Functions

5. A
6. D
7. C
8. D
9. E
10. C
11. D
12. A
13. E
14. A

## Level 1: Geometry

15. E
16. B
17. D
18. E
19. B
20. C
21. C
22. D
23. E
24. E

## Level 1: Probability and Statistics

25. E
26. D
27. C
28. D

## Level 1: Trigonometry

29. E
30. C
31. B
32. D

## Level 2: Number Theory

33. E
34. B
35. D
36. D

## Level 2: Algebra and Functions

37. C
38. E
39. D
40. C
41. D
42. B
43. C
44. E
45. A
46. D

## Level 2: Geometry

47. E
48. D
49. A
50. E
51. B
52. D
53. C
54. B
55. D
56. D

## Level 2: Probability and Statistics

57. B
58. C
59. A
60. D

## Level 2: Trigonometry

61. B
62. A
63. A
64. C

## Level 3: Number Theory

65. D
66. A
67. B
68. E

## Level 3: Algebra and Functions

69. E
70. C
71. B
72. D
73. C
74. A
75. D
76. C
77. C
78. D

## Level 3: Geometry

79. B
80. C
81. E
82. E
83. E
84. C
85. E

86. B
87. D
88. B

## Level 3: Probability and Statistics

89. B
90. A
91. D
92. C

## Level 3: Trigonometry

93. D
94. A
95. B
96. A

## Level 4: Number Theory

97. D
98. A
99. B
100. D

## Level 4: Algebra and Functions

101. D
102. C
103. D
104. A
105. A
106. E
107. C
108. B
109. E
110. D

## Level 4: Geometry

111. A
112. B
113. C
114. D

115. A
116. B
117. E
118. E
119. E
120. B

## LEVEL 4: PROBABILITY AND STATISTICS

121. B
122. B
123. C
124. D

## LEVEL 4: TRIGONOMETRY

125. A
126. E
127. E
128. C

## LEVEL 5: NUMBER THEORY

129. C
130. A
131. B
132. D

## LEVEL 5: ALGEBRA AND FUNCTIONS

133. E
134. C
135. C
136. D
137. C
138. C
139. B
140. A
141. B
142. B

# Level 5: Geometry

143. D
144. D
145. E
146. E
147. D
148. D
149. B
150. C
151. C
152. C

# Level 5: Probability and Statistics

153. E
154. E
155. A
156. D

# Level 5: Trigonometry

157. E
158. D
159. C
160. E

AFTERWORD

# YOUR ROAD TO SUCCESS

Congratulations! By practicing the problems in this book you have given yourself a significant boost to your SAT Math Level 1 Subject Test score. Go ahead and take a practice test. The math score you get should be much higher than the score you received on your last practice test.

What should you do to get your score even higher? Good news! You can use this book over and over again to continue to increase your score – right up to an 800. All you need to do is change the problems you are focusing on.

For each of the five subject areas go back and focus on problems that are right at and slightly above your current ability level. For example, if you have gotten all the Level 2 Geometry questions right, but you are still getting a few Level 3 Geometry questions wrong, then focus on Level 3 and 4 Geometry problems. Do this independently for each subject area.

Upon your next reading, try to solve each problem that you attempt in up to four different ways

- Using an SAT specific math strategy.
- The quickest way you can think of.
- The way you would do it in school.
- The easiest way for you.

Remember – the actual answer is not very important. What is important is to learn as many techniques as possible. This is the best way to simultaneously increase your current score, and increase your level of mathematical maturity. Keep doing problems from this book for about twenty minutes each day right up until two days before the test. Mark off the ones you get wrong and attempt them over and over again each week until you can get them right on your own.

I really want to thank you for putting your trust in me and my materials, and I want to assure you that you have made excellent use of your time by studying with this book. I wish you the best of luck on the SAT Math Subject Test, on getting into your choice college, and in life.

**Dr. Steve Warner**
**steve@SATPrepGet800.com**

# Actions to Complete After You Have Read This Book

1. **Take another practice test**

   You should see a substantial improvement in your score.

2. **Continue to practice SAT math subject test problems for 10 to 20 minutes each day**

   Keep practicing problems of the appropriate levels until two days before the test.

3. **Use my Facebook page for additional help**

   If you feel you need extra help that you cannot get from this book, please feel free to post your questions on my Facebook wall at www.facebook.com/SATPrepGet800.

4. **Review this book**

   If this book helped you, please post your positive feedback on the site you purchased it from; e.g. Amazon, Barnes and Noble, etc.

5. **Visit my website www.SATPrepGet800.com**

   You will find free content here that is updated weekly to help with your SAT preparation.

6. **Follow me on twitter**

www.twitter.com/SATPrepGet800

# About the Author

Steve Warner, a New York native, earned his Ph.D. at Rutgers University in Pure Mathematics in May, 2001. While a graduate student, Dr. Warner won the TA Teaching Excellence Award.

After Rutgers, Dr. Warner joined the Penn State Mathematics Department as an Assistant Professor. In September, 2002, Dr. Warner returned to New York to accept an Assistant Professor position at Hofstra University. By September 2007, Dr. Warner had received tenure and was promoted to Associate Professor. He has taught undergraduate and graduate courses in Precalculus, Calculus, Linear Algebra, Differential Equations, Mathematical Logic, Set Theory and Abstract Algebra.

Over that time, Dr. Warner participated in a five year NSF grant, "The MSTP Project," to study and improve mathematics and science curriculum in poorly performing junior high schools. He also published several articles in scholarly journals, specifically on Mathematical Logic.

Dr. Warner has over 15 years of experience in general math tutoring and over 10 years of experience in SAT math tutoring. He has tutored students both individually and in group settings.

In February, 2010 Dr. Warner released his first SAT prep book "The 32 Most Effective SAT Math Strategies." The second edition of this book was released in January, 2011. In February, 2012 Dr. Warner released his second SAT prep book "320 SAT Math Problems arranged by Topic and Difficulty Level." Between September 2012 and January 2013 Dr. Warner released his three book series "28 SAT Math Lessons to Improve Your Score in One Month." In June, 2013 Dr. Warner released the "SAT Prep Official Study Guide Math Companion." In November, 2013 Dr. Warner released the "ACT Prep Red Book – 320 Math Problems With Solutions," and in May, 2014 Dr. Warner released "320 SAT Math Problems arranged by Topic and Difficulty Level" for the Level 2 test.

Currently Dr. Warner lives in Manhattan with his two cats, Achilles and Odin. Since the age of 4, Dr. Warner has enjoyed playing the piano—especially compositions of Chopin as well as writing his own music. He also maintains his physical fitness through weightlifting.

# BOOKS BY DR. STEVE WARNER

24234883R00113

Made in the USA
Middletown, DE
18 September 2015